Mechanical Systems and Materials Six-Minute Problems

Second Edition

Harriet G. Cooke, PE

PPI2PASS.COM

Professional Publications, Inc. • Belmont, California

Benefit by Registering This Book with PPI

- Get book updates and corrections.
- Hear the latest exam news.
- Obtain exclusive exam tips and strategies.
- Receive special discounts.

Register your book at **ppi2pass.com/register**.

Report Errors and View Corrections for This Book

PPI is grateful to every reader who notifies us of a possible error. Your feedback allows us to improve the quality and accuracy of our products. You can report errata and view corrections at **ppi2pass.com/errata**.

MECHANICAL SYSTEMS AND MATERIALS SIX-MINUTE PROBLEMS
Second Edition

Current release of this edition: 7

Release History

date	edition number	revision number	update
Nov 2013	2	5	Minor corrections.
Mar 2015	2	6	Minor corrections. Minor cover updates.
Jul 2017	2	7	Minor corrections. Minor cover updates. Minor formatting and pagination updates.

© 2017 Professional Publications, Inc. All rights reserved.

All content is copyrighted by Professional Publications, Inc. (PPI). No part, either text or image, may be used for any purpose other than personal use. Reproduction, modification, storage in a retrieval system or retransmission, in any form or by any means, electronic, mechanical, or otherwise, for reasons other than personal use, without prior written permission from the publisher is strictly prohibited. For written permission, contact PPI at permissions@ppi2pass.com.

Printed in the United States of America.

PPI
1250 Fifth Avenue, Belmont, CA 94002
(650) 593-9119
ppi2pass.com

ISBN: 978-1-59126-551-1

Library of Congress Control Number: 2008929116

F E D C B A

Table of Contents

ABOUT THE AUTHOR v

NOMENCLATURE vii

TOPIC I: Breadth Problems
    Breadth Problems 1-1
    Breadth Problems Solutions 1-6

TOPIC II: Depth Problems
    Depth Problems 2-1
    Depth Problems Solutions 2-18

About the Author

Harriet G. Cooke received her professional engineering license in North Carolina in 1998. She holds three patents and a number of trade secrets, and has written a series of technical papers. Her interest in machine design began in grade school as she watched her father and grandfather design custom machinery for their woodworking business. In 1986, she began her engineering career as a cooperative student in the general engineering department of Georgia Power Company. As her experience increased, she moved to the general repair shop of the utility, where she designed her first large-scale production machine.

Ms. Cooke received her bachelor's degree in mechanical engineering from the Georgia Institute of Technology in 1991. She also participated in a special senior design project sponsored by NASA that was exhibited at the Lewis Research Center in Cleveland, Ohio, in the same year, and was later displayed at Sci-Trek in Atlanta, Georgia. Upon graduation, she moved to Hickory, North Carolina, to work for Siecor Corporation as a process development engineer in the research, development, and engineering facility. At Siecor, she was a member of the design team that created what was at that time the world's largest fiber optic cabling machine.

In 1998, she began a private practice as a consultant in mechanical design where she works on projects that are too small or unique for general purpose engineering firms.

Ms. Cooke is a member of the National Society of Professional Engineers. She is a past president of the local chapter of Professional Engineers of North Carolina, a former director of the chapter, and a former governor of the state association. In 2001, she received the President's Award by the state association. The local chapter also honored her as the 2001 Young Engineer of the Year.

Nomenclature

a	acceleration	ft/sec^2	m/s^2
a	relation constant	–	–
a	semimajor distance (of an ellipse)	ft	m
A	annual cost	$	$
A	area	ft^2	m^2
A	atomic weight	lbm/lbmol	kg/kmol
b	semiminor distance (of an ellipse)	ft	m
b	width	ft	m
BHP	brake horsepower	hp	hp
c	clearance	ft	m
c	parameter of the catenary	ft	m
c	specific heat	Btu/lbm-°F	J/kg·°C
C	circumference	ft	m
C	end resistent coefficient	–	–
C	spring index	–	–
C	thermal capacity rate	Btu/sec-°F	W/°C
d	diameter	ft	m
d	distance	ft	m
DP	degree of polymerization	–	–
e	eccentricity	ft	m
E	joint efficiency	–	–
E	modulus of elasticity	lbf/ft^2	Pa
EUAC	equivalent uniform annual cost	$	$
f	coefficient of friction	–	–
f	frequency	Hz	Hz
F	force or load	lbf	N
F	future value	$	$
F_R	applied radial load	lbf	N
F_T	applied thrust load	lbf	N
FS	factor of safety	–	–
g	acceleration of gravity	ft/sec^2	m/s^2
g_c	gravitational constant	ft-lbm/lbf-sec^2	–
G	shear modulus	lbf/ft^2	Pa
G	shock transmission	g	g
h	head, head loss, or height	ft	m
H	generated heat	Btu/hr	W
HHV	higher heating value	Btu/lbm	J/kg
I	moment of inertia	ft^4	m^4
J	polar moment of inertia	ft^4	m^4
k	radius of gyration	ft	m
k	spring constant	lbf/ft	N/m
k	stress concentration factor	–	–
k_r	torsional spring constant	ft-lbf/rad	N·m/rad
K	factor or ratio	–	–
L	lead	ft	m
L	length	ft	m
L	life	hr	h
m	mass	lbm	kg
$\dot{m}$	mass flow rate	lbm/sec	kg/s
M	moment	ft-lbf	N·m
MC	maintenance cost or material cost	$	$
MCR	material consumption rate	lbm/hr	kg/h
MF	magnification factor	–	–
MR	modulus of resilience	lbf/ft^2	Pa
MRC	mean repair cost	$	$
MTBF	mean time between failures	hr	h
MTTF	mean time to failure	hr	h
MTTR	mean time to repair	hr	h
MW	molecular weight	lbm/lbmol	kg/kmol
n	number of units	–	–
n	rotational speed	rpm	rpm
N	endurance life	–	–
N	normal force	lbf	N
N	number of units	–	–
p	circular pitch	ft	m
p	pressure	lbf/ft^2	Pa
P	equivalent radial load	lbf	N
P	power	hp	W
P	present value	$	$
PC	process cost	$	$
Q	heat flow rate	Btu/hr	W
r	radius	ft	m

Symbol	Description	US Units	SI Units
r	ratio of forcing to natural frequency	–	–
R	range	ft	m
R	rate of corrosion loss	ft/yr	m/yr
R	reaction force	lbf	N
R	reliability	–	–
R	specific gas constant	ft-lbf/lbm-°R	J/kg·K
R^*	universal gas constant	ft-lbf/lbmol-°R	J/kmol·K
S	allowable stress	lbf/ft²	Pa
S	sag	ft	m
S	section modulus	ft³	m³
S	strength	lbf/ft²	Pa
t	thickness	ft	m
t	time	sec	s
t_e	effective throat size	ft	m
T	flight time	sec	s
T	temperature	°F	°C
T	tension or thrust force	lbf	N
T	torque	ft-lbf	N·m
TR	transmissibility	–	–
TV	train value	–	–
U_R	modulus of resilience	lbf/ft²	Pa
v	velocity	ft/sec	m/s
V	rotation factor	–	–
V	vertical shear force	lbf	N
V	volume	ft³	m³
w	weight per unit of length	lbf/ft	N/m
W	weight	lbf	–
W	work	ft-lbf	J
WHP	water horsepower	hp	hp
x	horizontal distance or position	ft	m
X	radial load factor	–	–
y	vertical distance or position	ft	m
y	weld size	ft	m
Y	thrust load factor	–	–

Symbols

Symbol	Description	US Units	SI Units
α	angle	deg	deg
α	angular acceleration	deg/sec²	deg/s²
α	coefficient of linear thermal expansion	1/°F	1/°C
γ	pitch angle	deg	deg
γ	specific weight	lbf/ft³	–
δ	deformation or deflection	ft	m
ϵ	true strain	ft/ft	m/m
ζ	damping ratio	–	–
η	efficiency	–	–
θ	angle	deg	deg
θ	angular position	deg	deg
κ	torsion constant	ft⁴/deg	m⁴/deg
μ	absolute viscosity	lbf-sec/ft²	Pa·s
ν	Poisson's ratio	–	–
ρ	density	lbm/ft³	kg/m³
σ	stress	lbf/ft²	Pa
σ_1	maximum principal stress	lbf/ft²	Pa
σ_2	minimum principal stress	lbf/ft²	Pa
τ	shear stress	lbf/ft²	Pa
ω	angular frequency	Hz	Hz
ω	angular velocity	deg/sec	deg/s

Subscripts

0	initial
a	allowable, amplitude
A	added
alt	alternating
b	bearing, bending
B	Bergsträsser
c	center, compression, compressive
C	carbon
d	damaged
D	design
e	endurance
eq	equivalent
f	failure, fatigue, flow, forcing
H	high
H	hydrogen
hr	per hour
i	inner, inside
L	low
lbm	per pound
m	machine, mean
max	maximum
min	minimum
mod	modified
n	natural
o	outer, outside
O	oxygen
op	operating
p	constant pressure
r	range, rivet
R	radial, rated
s	shear, sheet

S	stress
st	static
t	tangential, tensile, tension, torsional, transmitted
T	thrust
u	ultimate
v	vertical
y	yield

1 Breadth Problems

MECHANICAL SYSTEMS AND MATERIALS

PROBLEM 1

A 0.25 in diameter steel support wire with modulus of elasticity 30×10^6 psi is normally strung between two utility poles 125 ft apart under 2500 lbf of tension. A winch maintains the tension by stretching the wire. A new procedure requires 6250 lbf of tension. The length of additional wire the winch must decrease to reach the new tension is most nearly

(A) 2.6 in
(B) 3.8 in
(C) 6.4 in
(D) 8.9 in

Hint: Find the wire length for each tension.

PROBLEM 2

A bridge's steel midspan is to be installed at 35°F during the winter. Rivet holes at each end of the midspan must measure 48.5 ft from the center. The I-beams for the midspan are prepared in a fabrication shop with a controlled temperature of 72°F. The change to the center distance that should be made when the rivet holes are drilled in the shop is most nearly

(A) −0.14 in
(B) −0.012 in
(C) 0.012 in
(D) 0.14 in

Hint: Use the coefficient of linear thermal expansion of steel.

PROBLEM 3

A 250 hp motor turns a shaft at 2400 rpm. The shaft steps down in diameter from 3.0 in to 2.5 in.

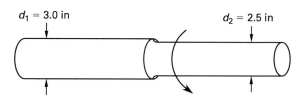

The maximum torsional shear stress is most nearly

(A) 1700 psi
(B) 2100 psi
(C) 2800 psi
(D) 3700 psi

Hint: Find the stress concentration factor for a filleted shaft in torsion.

PROBLEM 4

The principal stresses on the object shown are most nearly

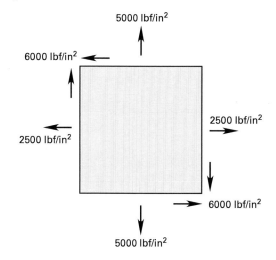

(A) 2400 psi, −9900 psi
(B) 2500 psi, 5000 psi
(C) 9900 psi, −2400 psi
(D) 3800 psi, 6100 psi

Hint: Determine the applied stresses, σ_x, σ_y, and τ. Remember that tensile stresses are positive.

PROBLEM 5

A flywheel and gear are located on a stationary shaft supported by two bearings as shown. The flywheel applies a 200 lbf vertical load, and the gear applies a 50 lbf vertical load.

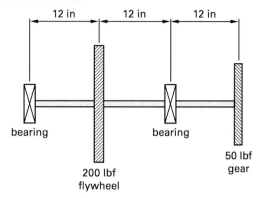

At what locations on the shaft, as measured from the left end, are the shear forces undefined?

(A) 0 in, 12 in, 24 in, and 36 in

(B) 0 in, 22 in, and 36 in

(C) 12 in and 24 in

(D) 22 in

Hint: Draw the shear diagram of the shaft.

PROBLEM 6

A cube with 2 in edges is made from a brittle nickel alloy (ultimate tensile strength of 35,000 psi, ultimate compressive strength of 125,000 psi). In use, the cube experiences combined forces of 56,000 lbf (tensile), 100,000 lbf (compressive), and 10,000 psi (shear). A factor of safety greater than 2.0 is required. What is the failure mode for this material?

(A) compression

(B) principal stresses

(C) tension

(D) shear

Hint: Construct a diagram of the region of acceptable design using the modified Mohr theory.

PROBLEM 7

A soft, steel block experiences a 15,000 psi horizontal compressive stress, a 10,000 psi vertical compressive stress, and a shear stress of 5000 psi as shown. The modulus of elasticity is 29×10^6 psi, and the yield strength is 48,000 psi. Poisson's ratio is 0.27.

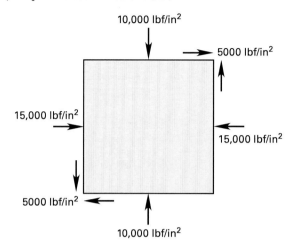

Using strain energy theory, the factor of safety and the modulus of resilience are found to be most nearly

(A) FS = 2.7, MR = 5.8 psi

(B) FS = 2.7, MR = 40 psi

(C) FS = 2.8, MR = 40 psi

(D) FS = 3.1, MR = 40 psi

Hint: Find the principal stresses and solve for the yield strength.

PROBLEM 8

A fixture experiences a static stress of 3500 psi and a dynamic stress of 4000 psi. The yield strength is 43,000 psi, and the endurance strength is 8820 psi. The fatigue stress concentration factor is 1.2. The equivalent stress is most nearly

(A) 2800 psi

(B) 3200 psi

(C) 3400 psi

(D) 3500 psi

Hint: Use the mean and alternating stresses to calculate the equivalent stress.

PROBLEM 9

A 200 ft cable weighs 10 lbf/ft.

The tension needed at point D to maintain a minimum sag of 50 ft is most nearly

(A) 500 lbf

(B) 800 lbf

(C) 1300 lbf

(D) 4300 lbf

Hint: Determine the parameter of the catenary.

PROBLEM 10

Truss 1 is modified by removing members BG, BH, CH, DJ, EJ, and EK to become truss 2. Five 2000 lbf loads had been applied to truss 1 at 30 ft intervals as shown; the loads are now consolidated and centered on truss 2.

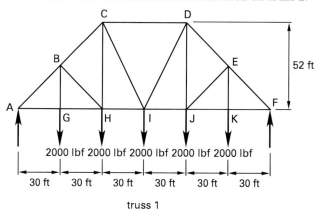

truss 1

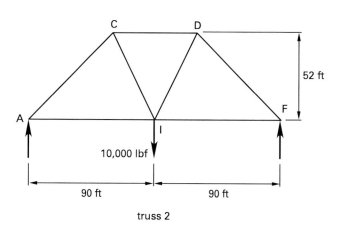

truss 2

The percentage of change in tension in member CD is most nearly

(A) −30%

(B) 0%

(C) 20%

(D) 70%

Hint: Use the cut-and-sum method and the method of sections.

PROBLEM 11

A part hangs 10 ft above the floor, suspended by a clip moving at 100 ft/min on an overhead conveyor. A trip opens the clip, which allows the part to fall into a bin 2 ft above the floor. The horizontal distance from trip to bin is most nearly

(A) 14 in

(B) 16 in

(C) 35 in

(D) 190 in

Hint: This process is a demonstration of horizontal projection.

PROBLEM 12

A mechanism starts from rest and accelerates at 50 rad/sec^2. After 10 sec, a bolt flies off. The bolt circle has a diameter of 12 in. The speed of the bolt when it leaves the mechanism is most nearly

(A) 54 mph

(B) 85 mph

(C) 170 mph

(D) 1100 mph

Hint: Calculate the angular velocity and the circumference of the bolt circle.

PROBLEM 13

The slider crank assembly shown begins at rest in a straight configuration. The crank rotates clockwise and accelerates at 1.75 rad/sec².

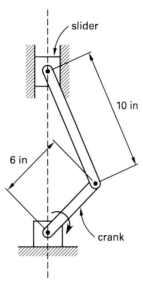

At 10 sec, the velocity of the slider is most nearly

(A) 0.58 ft/sec

(B) 2.9 ft/sec

(C) 5.8 ft/sec

(D) 36 ft/sec

Hint: Use the crank's angular velocity to determine its position at 10 sec.

PROBLEM 14

A fillet weld connects two metal plates as shown. The plates are 0.50 in thick and 10 in wide. The weld size is 0.375 in. A 22,500 lbf force is applied.

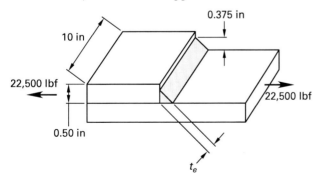

The assumed shear carried by the weld is most nearly

(A) 2500 psi

(B) 4500 psi

(C) 6000 psi

(D) 8500 psi

Hint: Start by calculating the effective throat size.

PROBLEM 15

A 5 lbm mass hanging from the end of a bar is used as a trip as shown. The mass of the bar and the friction of the pivot are inconsequential and may be disregarded. Slides hit the mass at a velocity of 61.5 ft/min. The bar is 2 ft long, and the spring constant is 10 lbf/in.

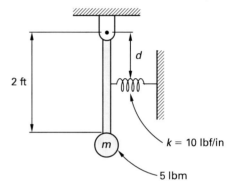

The distance, d, needed between the pivot point and the spring for the mass, m, to reach a maximum horizontal amplitude, A, of 3 in is most nearly

(A) 0.69 in

(B) 0.73 in

(C) 3.4 in

(D) 4.1 in

Hint: This configuration is a vertical constrained compound pendulum. Begin by determining the angular velocity of the system.

PROBLEM 16

A hemispherical pressure head is designed to operate at 175 psig and 500°F. The inside diameter is 50 in. The head is constructed of SA-515 65 plate with a fully examined single butt weld with an integral backing strip. Assuming that economy is a concern, what standard plate thickness should be specified for use?

(A) 0.125 in

(B) 0.188 in

(C) 0.313 in

(D) 0.750 in

Hint: Start by looking up the values for SA-515 65 plate and fully examined single butt welds.

PROBLEM 17

A welded pressure vessel is shown.

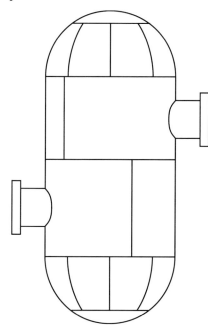

According to the ASME boiler and pressure vessel code, how many of the 17 welds visible here are Category A?

(A) 8

(B) 12

(C) 13

(D) 17

Hint: Review the definitions for pressure vessel welding categories.

PROBLEM 18

When two dissimilar materials are coupled

    I. the more electronegative material becomes the anode

   II. the more electronegative material will dissolve

  III. an electrolytic cell is formed

  IV. the cathode is sacrificed

(A) I and III only

(B) II and IV only

(C) I, II, and III

(D) I, II, and IV

Hint: Review the definitions of anode, cathode, and electrolytic cell.

PROBLEM 19

A nickel plate is tested in a corrosive environment. The density of the nickel is 8602 kg/m³. A 5 cm² sample is exposed for 72 hours and loses 1.5 g. The predicted material loss for the nickel plate in use is most nearly

(A) 0.0424 μm/yr

(B) 0.484 μm/yr

(C) 42.4 μm/yr

(D) 42,400 μm/yr

Hint: The rate of corrosion loss is the ratio of material lost over time. The materials section in a reference book will give the necessary formula and constants.

SOLUTION 1

The cross-sectional area of the wire is

$$A = \pi r^2 = \pi \left(\frac{0.25 \text{ in}}{2}\right)^2$$
$$= 0.049 \text{ in}^2$$

The modulus of elasticity for steel is

$$E = 30 \times 10^6 \text{ psi}$$

The change in wire length for the old tension is

$$\delta_1 = \frac{L_o F_1}{EA} = \frac{(125 \text{ ft})\left(12 \frac{\text{in}}{\text{ft}}\right)(2500 \text{ lbf})}{\left(30 \times 10^6 \frac{\text{lbf}}{\text{in}^2}\right)(0.049 \text{ in}^2)}$$
$$= 2.55 \text{ in}$$

The change in wire length for the new tension is

$$\delta_2 = \frac{L_o F_2}{EA}$$
$$= \frac{(125 \text{ ft})\left(12 \frac{\text{in}}{\text{ft}}\right)(6250 \text{ lbf})}{\left(30 \times 10^6 \frac{\text{lbf}}{\text{in}^2}\right)(0.049 \text{ in}^2)}$$
$$= 6.38 \text{ in}$$

The additional change in length from the old tension to the new tension is

$$\delta = \delta_2 - \delta_1 = 6.38 \text{ in} - 2.55 \text{ in}$$
$$= 3.83 \text{ in} \quad (3.8 \text{ in})$$

The winch would have to pull 3.8 in of additional wire to achieve the new tension.

The answer is (B).

Why Other Options Are Wrong

(A) This incorrect solution is the change in length for the old tension.

(C) This incorrect solution is the change in length for the new tension.

(D) This incorrect solution results when the changes in wire length for the old and new tensions are added together instead of subtracted.

SOLUTION 2

From a table of steel properties, the coefficient of linear thermal expansion of steel is

$$\alpha = 6.5 \times 10^{-6} \text{ 1/°F}$$

The change in the length of the beam from the cold temperature to the hot temperature is

$$\Delta L = \alpha L_0 (T_2 - T_1)$$
$$= \left(6.5 \times 10^{-6} \frac{1}{\text{°F}}\right)(48.5 \text{ ft})\left(12 \frac{\text{in}}{\text{ft}}\right)(72\text{°F} - 35\text{°F})$$
$$= 0.140 \text{ in}$$

The answer is (D).

Why Other Options Are Wrong

(A) This incorrect solution results when the temperature change is reversed.

(B) This incorrect solution results when the change in length is not converted to inches and the temperature change is reversed.

(C) This incorrect solution results when the change in length is not converted to inches.

SOLUTION 3

The shoulder has a radius of

$$r = \frac{d_1}{2} - \frac{d_2}{2}$$
$$= \frac{3.0 \text{ in}}{2} - \frac{2.5 \text{ in}}{2}$$
$$= 0.25 \text{ in}$$

Find the stress concentration factor for the fillet.

$$\frac{r}{d_2} = \frac{0.25 \text{ in}}{2.5 \text{ in}} = 0.1$$
$$\frac{d_1}{d_2} = \frac{3 \text{ in}}{2.5 \text{ in}} = 1.2$$

Find (0.1, 1.2) on a graph of stress concentration factors for a filleted shaft in torsion.

$$k = 1.3$$

Using the smaller diameter of the shaft, calculate the polar moment of inertia.

$$J = \left(\frac{\pi}{2}\right)\left(\frac{d_2}{2}\right)^4$$
$$= \left(\frac{\pi}{2}\right)\left(\frac{2.5 \text{ in}}{2}\right)^4$$
$$= 3.835 \text{ in}^4$$

The torque generated by the motor is given by the formula

$$T_{\text{in-lbf}} = \frac{63{,}025 P_{\text{hp}}}{n_{\text{rpm}}}$$
$$= \frac{(63{,}025)(250 \text{ hp})}{2400 \ \frac{\text{rev}}{\text{min}}}$$
$$= 6565 \text{ in-lbf}$$

The torsional shear stress with the stress concentration factor applied is

$$\tau = k\left(\frac{T\left(\frac{d_2}{2}\right)}{J}\right)$$
$$= (1.3)\left(\frac{(6565 \text{ in-lbf})\left(\frac{2.50 \text{ in}}{2}\right)}{3.835 \text{ in}^4}\right)$$
$$= 2782 \text{ psi} \quad (2800 \text{ psi})$$

The answer is (C).

Why Other Options Are Wrong

(A) This incorrect solution results when the stress concentration factor is misapplied.

(B) This incorrect solution results when the stress concentration factor is neglected.

(D) This incorrect solution results when the wrong chart is used to find the stress concentration factor.

SOLUTION 4

Construct Mohr's circle. Calculate σ_c to plot the center.

$$\sigma_c = \frac{1}{2}(\sigma_x + \sigma_y)$$
$$= \left(\frac{1}{2}\right)\left(2500 \ \frac{\text{lbf}}{\text{in}^2} + 5000 \ \frac{\text{lbf}}{\text{in}^2}\right)$$
$$= 3750 \text{ lbf/in}^2$$

Calculate the radius of the circle, r, around point σ_c.

$$r = \sqrt{\frac{1}{4}(\sigma_x - \sigma_y)^2 + \tau^2}$$
$$= \sqrt{\left(\frac{1}{4}\right)\left(2500 \ \frac{\text{lbf}}{\text{in}^2} - 5000 \ \frac{\text{lbf}}{\text{in}^2}\right)^2 + \left(6000 \ \frac{\text{lbf}}{\text{in}^2}\right)^2}$$
$$= 6128.8 \text{ lbf/in}^2$$

Draw the circle using σ_c and r.

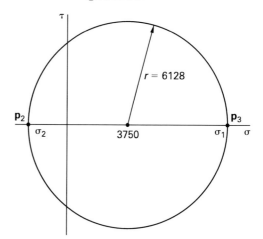

The larger principal stress, σ_1, is at point $\mathbf{p}_3$ on Mohr's circle. The smaller principal stress, σ_2, is at point $\mathbf{p}_2$. Use the center point and circle radius to calculate these values.

$$\sigma_1 = 3750 \ \frac{\text{lbf}}{\text{in}^2} + 6128.8 \ \frac{\text{lbf}}{\text{in}^2}$$
$$= 9878.8 \text{ lbf/in}^2 \quad (9900 \text{ psi})$$
$$\sigma_2 = 3750 \ \frac{\text{lbf}}{\text{in}^2} - 6128.8 \ \frac{\text{lbf}}{\text{in}^2}$$
$$= -2378.8 \text{ lbf/in}^2 \quad (-2400 \text{ psi})$$

The answer is (C).

Why Other Options Are Wrong

(A) This solution results when the tensile normal stresses are incorrectly considered to be negative.

(B) This solution results when principal stresses are incorrectly determined by inspection.

(D) This incorrect solution results when σ_c and r are used as the principal stresses.

SOLUTION 5

Determine the reactions.

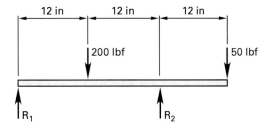

Calculate the sum of the forces in the y-axis.

$$\sum F_y = 0$$
$$R_1 + R_2 - 200 \text{ lbf} - 50 \text{ lbf} = 0$$
$$R_1 = 250 \text{ lbf} - R_2$$

Calculate the sum of the moments around the first reaction force, and use this to determine the values of the reaction forces.

$$\sum M_{R_1} = 0$$

$$(12 \text{ in})(200 \text{ lbf}) - (24 \text{ in})(R_2) + (36 \text{ in})(50 \text{ lbf}) = 0$$

$$R_2 = \frac{(12 \text{ in})(200 \text{ lbf}) + (36 \text{ in})(50 \text{ lbf})}{24 \text{ in}}$$
$$= 175 \text{ lbf}$$
$$R_1 = 250 \text{ lbf} - R_2$$
$$= 75 \text{ lbf}$$

Plot the shear diagram.

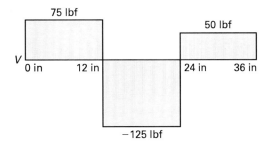

At 0 in, 12 in, 24 in, and 36 in, the shear force is a vertical line and is undefined.

The answer is (A).

Why Other Options Are Wrong

(B) This incorrect solution results when the moment diagram is considered instead of the shear diagram.

(C) This incorrect solution results when the ends of the shaft are neglected.

(D) This incorrect solution results when the moment diagram is considered instead of the shear diagram and the ends of the shaft are neglected.

SOLUTION 6

The area of one face of the cube is

$$A = bh = (2 \text{ in})(2 \text{ in})$$
$$= 4 \text{ in}^2$$

The tensile stress is

$$\sigma_t = \frac{F_t}{A} = \frac{56{,}000 \text{ lbf}}{4 \text{ in}^2}$$
$$= 14{,}000 \text{ lbf/in}^2$$

The factor of safety for tension is

$$\text{FS}_t = \frac{S_t}{\sigma_t}$$
$$= \frac{35{,}000 \ \dfrac{\text{lbf}}{\text{in}^2}}{14{,}000 \ \dfrac{\text{lbf}}{\text{in}^2}}$$
$$= 2.5$$

The compressive stress is

$$\sigma_c = \frac{F_c}{A} = \frac{-100{,}000 \text{ lbf}}{4 \text{ in}^2}$$
$$= -25{,}000 \text{ lbf/in}^2$$

The factor of safety of compression is

$$\text{FS}_c = \frac{S_c}{\sigma_c}$$
$$= \frac{-125{,}000 \ \dfrac{\text{lbf}}{\text{in}^2}}{-25{,}000 \ \dfrac{\text{lbf}}{\text{in}^2}}$$
$$= 5.0$$

The compressive and tensile stresses both pass the factor of safety requirements.

Find the region of acceptable design.

$$\sigma_t = \frac{S_t}{\text{FS}}$$
$$= \frac{35{,}000 \ \frac{\text{lbf}}{\text{in}^2}}{2.0}$$
$$= 17{,}500 \ \text{lbf/in}^2$$

$$\sigma_c = \frac{S_c}{\text{FS}}$$
$$= \frac{-125{,}000 \ \frac{\text{lbf}}{\text{in}^2}}{2.0}$$
$$= -62{,}500 \ \text{lbf/in}^2$$

The maximum shear stress is

$$\tau_{\max} = \tfrac{1}{2}\sqrt{(\sigma_c - \sigma_t)^2 + (2\tau)^2}$$
$$= \left(\tfrac{1}{2}\right)\sqrt{\left(-25{,}000 \ \frac{\text{lbf}}{\text{in}^2} - 14{,}000 \ \frac{\text{lbf}}{\text{in}^2}\right)^2 + \left((2)\left(10{,}000 \ \frac{\text{lbf}}{\text{in}^2}\right)\right)^2}$$
$$= 21{,}915 \ \text{lbf/in}^2$$

The maximum principal stress is

$$\sigma_1 = \tfrac{1}{2}(\sigma_c + \sigma_t) + \tau_{\max}$$
$$= \left(\tfrac{1}{2}\right)\left(-25{,}000 \ \frac{\text{lbf}}{\text{in}^2} + 14{,}000 \ \frac{\text{lbf}}{\text{in}^2}\right) + 21{,}915 \ \frac{\text{lbf}}{\text{in}^2}$$
$$= 16{,}415 \ \text{lbf/in}^2$$

The maximum factor of safety is

$$\text{FS}_1 = \frac{S_t}{\sigma_1} = \frac{35{,}000 \ \frac{\text{lbf}}{\text{in}^2}}{16{,}415 \ \frac{\text{lbf}}{\text{in}^2}}$$
$$= 2.1$$

The minimum principal stress is

$$\sigma_2 = \tfrac{1}{2}(\sigma_c + \sigma_t) - \tau_{\max}$$
$$= \left(\tfrac{1}{2}\right)\left(-25{,}000 \ \frac{\text{lbf}}{\text{in}^2} + 14{,}000 \ \frac{\text{lbf}}{\text{in}^2}\right) - 21{,}915 \ \frac{\text{lbf}}{\text{in}^2}$$
$$= -27{,}415 \ \text{lbf/in}^2$$

The minimum factor of safety is

$$\text{FS}_2 = \frac{S_t}{|\sigma_2|} = \frac{35{,}000 \ \frac{\text{lbf}}{\text{in}^2}}{27{,}415 \ \frac{\text{lbf}}{\text{in}^2}}$$
$$= 1.3$$

Plot the acceptable design region and the point (σ_1, σ_2).

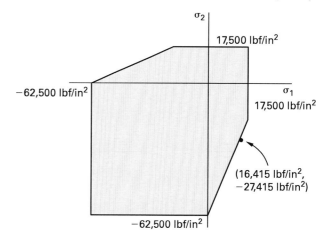

The point lies outside the acceptable design region and falls in quadrant IV. The design will fail in shear.

The answer is (D).

Why Other Options Are Wrong

(A) This incorrect solution results when the failure mode is determined by assuming the largest force will be first to fail.

(B) This solution is incorrect because the principal stresses are not a failure mode. They are calculated to determine whether the design will fail due to the shear stress applied.

(C) This incorrect solution results when the factors of safety for the tensile and compressive forces are compared and shear is neglected.

SOLUTION 7

The maximum shear stress is

$$\tau_{\max} = \tfrac{1}{2}\sqrt{(\sigma_x - \sigma_y)^2 + (2\tau)^2}$$
$$= \left(\tfrac{1}{2}\right)\sqrt{\left(-15{,}000 \ \frac{\text{lbf}}{\text{in}^2} - \left(-10{,}000 \ \frac{\text{lbf}}{\text{in}^2}\right)\right)^2 + \left((2)\left(5000 \ \frac{\text{lbf}}{\text{in}^2}\right)\right)^2}$$
$$= 5590 \ \text{lbf/in}^2$$

The maximum principal stress is

$$\sigma_1 = \tfrac{1}{2}(\sigma_x + \sigma_y) + \tau_{\max}$$

$$= \left(\tfrac{1}{2}\right)\left(-15{,}000 \ \frac{\text{lbf}}{\text{in}^2} + \left(-10{,}000 \ \frac{\text{lbf}}{\text{in}^2}\right)\right) + 5590 \ \frac{\text{lbf}}{\text{in}^2}$$

$$= -6910 \ \text{lbf/in}^2$$

The minimum principal stress is

$$\sigma_2 = \tfrac{1}{2}(\sigma_x + \sigma_y) - \tau_1$$

$$= \left(\tfrac{1}{2}\right)\left(-15{,}000 \ \frac{\text{lbf}}{\text{in}^2} + \left(-10{,}000 \ \frac{\text{lbf}}{\text{in}^2}\right)\right)$$

$$- 5590 \ \frac{\text{lbf}}{\text{in}^2}$$

$$= -18{,}090 \ \text{lbf/in}^2$$

The factor of safety is

$$\text{FS} = \frac{S_y}{\sqrt{\sigma_1^2 + \sigma_2^2 - 2\nu\sigma_1\sigma_2}}$$

$$= \frac{48{,}000 \ \frac{\text{lbf}}{\text{in}^2}}{\sqrt{\left(-6910 \ \frac{\text{lbf}}{\text{in}^2}\right)^2 + \left(-18{,}090 \ \frac{\text{lbf}}{\text{in}^2}\right)^2 - (2)(0.27)\left(-6910 \ \frac{\text{lbf}}{\text{in}^2}\right)\left(-18{,}090 \ \frac{\text{lbf}}{\text{in}^2}\right)}}$$

$$= 2.737 \quad (2.7)$$

The modulus of resilience is

$$\text{MR} = \frac{S_y^2}{2E}$$

$$= \frac{\left(48{,}000 \ \frac{\text{lbf}}{\text{in}^2}\right)^2}{(2)\left(29 \times 10^6 \ \frac{\text{lbf}}{\text{in}^2}\right)}$$

$$= 39.72 \ \text{lbf/in}^2 \quad (40 \ \text{psi})$$

The answer is (B).

Why Other Options Are Wrong

(A) This incorrect solution results when the strain energy is calculated instead of the modulus of resilience.

(C) This incorrect solution results when the maximum shear is not calculated.

(D) This incorrect solution results when σ_x and σ_y are used instead of the principal stresses.

SOLUTION 8

Identify the static and dynamic stresses.

$$\text{static stress} = \sigma_{\min}$$
$$= 3500 \ \text{lbf/in}^2$$
$$\text{dynamic stress} = \sigma_r$$
$$= 4000 \ \text{lbf/in}^2$$

The maximum stress is

$$\sigma_{\max} = \sigma_r + \sigma_{\min}$$
$$= 4000 \ \frac{\text{lbf}}{\text{in}^2} + 3500 \ \frac{\text{lbf}}{\text{in}^2}$$
$$= 7500 \ \text{lbf/in}^2$$

The alternating stress is

$$\sigma_{\text{alt}} = \tfrac{1}{2}\sigma_r$$
$$= \left(\tfrac{1}{2}\right)\left(4000 \ \frac{\text{lbf}}{\text{in}^2}\right)$$
$$= 2000 \ \text{lbf/in}^2$$

The mean stress is

$$\sigma_m = \frac{\sigma_{\max} + \sigma_{\min}}{2}$$
$$= \frac{7500 \ \frac{\text{lbf}}{\text{in}^2} + 3500 \ \frac{\text{lbf}}{\text{in}^2}}{2}$$
$$= 5500 \ \text{lbf/in}^2$$

The equivalent stress is

$$\sigma_{\text{eq}} = k_f \sigma_{\text{alt}} + \left(\frac{S_e}{S_y}\right)\sigma_m$$

$$= (1.2)\left(2000 \ \frac{\text{lbf}}{\text{in}^2}\right) + \left(\frac{8820 \ \frac{\text{lbf}}{\text{in}^2}}{43{,}000 \ \frac{\text{lbf}}{\text{in}^2}}\right)\left(5500 \ \frac{\text{lbf}}{\text{in}^2}\right)$$

$$= 3528 \ \text{lbf/in}^2 \quad (3500 \ \text{psi})$$

The answer is (D).

Why Other Options Are Wrong

(A) This incorrect solution results when the stress concentration factor is misapplied to the alternating stress.

(B) This incorrect solution results when the dynamic stress is assumed to be σ_{max}.

(C) This incorrect solution results when the stress concentration factor is applied to the mean stress instead of to the alternating stress.

Why Other Options Are Wrong

(A) This incorrect solution results when the distance, y, is assumed to be equal to the cable sag.

(B) This incorrect solution results when the tension is calculated for the lowest part of the sag.

(D) This incorrect solution results when the full length of the cable, instead of half the length, is used for the calculation.

SOLUTION 9

The cable length for calculation is

$$L = \frac{200 \text{ ft}}{2} = 100 \text{ ft}$$

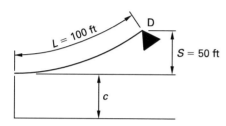

The parameter of the catenary, c, is calculated from the relationship

$$c + S = \sqrt{L^2 + c^2}$$
$$c + 50 \text{ ft} = \sqrt{(100 \text{ ft})^2 + c^2}$$
$$(c + 50 \text{ ft})^2 = 10{,}000 \text{ ft}^2 + c^2$$
$$c^2 + (100 \text{ ft})c + 2500 \text{ ft}^2 = 10{,}000 \text{ ft}^2 + c^2$$
$$(100 \text{ ft})c = 7500 \text{ ft}^2$$
$$c = 75 \text{ ft}$$

The total height to point D is

$$h = S + c = 50 \text{ ft} + 75 \text{ ft}$$
$$= 125 \text{ ft}$$

The tension of the cable at point D is

$$T_D = Wh = \left(10 \frac{\text{lbf}}{\text{ft}}\right)(125 \text{ ft})$$
$$= 1250 \text{ ft} \quad (1300 \text{ lbf})$$

The answer is (C).

SOLUTION 10

Find the reaction forces for truss 1.

$$\sum M_A = 0 \text{ ft-lbf}$$

$$F_G d_{AG} + F_H d_{AH} + F_I d_{AI}$$
$$+ F_J d_{AJ} + F_K d_{AK} - R_F d_{AF} = 0 \text{ ft-lbf}$$

$$(2000 \text{ lbf})(30 \text{ ft}) + (2000 \text{ lbf})(60 \text{ ft})$$
$$+ (2000 \text{ lbf})(90 \text{ ft}) + (2000 \text{ lbf})(120 \text{ ft})$$
$$+ (2000 \text{ lbf})(150 \text{ ft}) - R_F(180 \text{ ft}) = 0 \text{ ft-lbf}$$

$$R_F = 5000 \text{ lbf}$$
$$\sum F_y = 0 \text{ ft-lbf}$$
$$F_G + F_H + F_I + F_J + F_K - R_A - R_F = 0 \text{ ft-lbf}$$
$$(5)(2000 \text{ lbf}) - R_A - 5000 \text{ lbf} = 0 \text{ ft-lbf}$$
$$R_A = 5000 \text{ lbf}$$

Find the reaction forces for truss 2.

$$\sum M_A = 0 \text{ ft-lbf}$$
$$F_I d_{AI} - R_F d_{AF} = 0 \text{ ft-lbf}$$
$$(10{,}000 \text{ lbf})(90 \text{ ft}) - R_F(180 \text{ ft}) = 0 \text{ ft-lbf}$$
$$R_F = 5000 \text{ lbf}$$

$$\sum F_y = 0 \text{ ft-lbf}$$
$$F_I - R_A - R_F = 0 \text{ ft-lbf}$$
$$10{,}000 \text{ lbf} - R_A - 5000 \text{ lbf} = 0 \text{ ft-lbf}$$
$$R_A = 5000 \text{ lbf}$$

Use the method of sections to determine the force in member CD for truss 1.

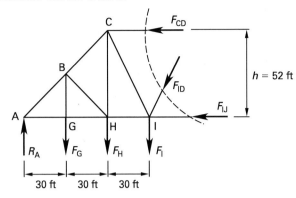

truss 1

$$\sum M_I = 0 \text{ ft-lbf}$$
$$-R_A d_{AI} + F_G d_{GI} + F_H d_{HI} + F_{CD} h = 0 \text{ ft-lbf}$$

$$-(5000 \text{ lbf})(90 \text{ ft}) + (2000 \text{ lbf})(60 \text{ ft})$$
$$+(2000 \text{ lbf})(30 \text{ ft}) + F_{CD}(52 \text{ ft}) = 0 \text{ ft-lbf}$$
$$F_{CD} = 5192.3 \text{ lbf}$$

Use the method of sections to determine the force in member CD for truss 2.

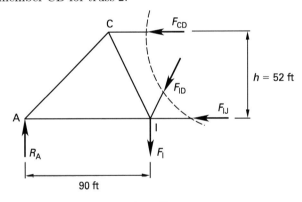

truss 2

$$\sum M_I = 0 \text{ ft-lbf}$$
$$-R_A d_{AI} + F_{CD} h = 0 \text{ ft-lbf}$$
$$-(5000 \text{ lbf})(90 \text{ ft}) - F_{CD}(52 \text{ ft}) = 0 \text{ ft-lbf}$$
$$F_{CD} = 8653.8 \text{ lbf}$$

The percentage of change from truss 1 to truss 2 is

$$\Delta\% = \left(\frac{F_{CD} \text{ for truss 2}}{F_{CD} \text{ for truss 1}} - 1\right) \times 100\%$$
$$= \left(\frac{8653.8 \text{ lbf}}{5192.3 \text{ lbf}} - 1\right) \times 100\%$$
$$= 66.67\% \quad (70\%)$$

The answer is (D).

Why Other Options Are Wrong

(A) This incorrect solution results when the wrong direction is chosen for the moments about I for truss 1.

(B) This solution results from the incorrect assumption that, since member CD is not touched or changed by removal of the old members and the sum of the forces is the same, the tension on member CD will not change.

(C) This incorrect solution results when the distance between one member and the next is used instead of the total distance from the moment location when calculating moment force for truss 2.

SOLUTION 11

The vertical distance the part falls from the clip to the bin is

$$y = h_1 - h_2$$
$$= 10 \text{ ft} - 2 \text{ ft}$$
$$= 8 \text{ ft}$$

The time it takes for the part to fall this distance is found from the formula

$$y = \frac{1}{2}gt^2$$
$$t = \sqrt{\frac{2y}{g}}$$
$$= \sqrt{\frac{(2)(8 \text{ ft})}{32.2 \frac{\text{ft}}{\text{sec}^2}}}$$
$$= 0.70 \text{ sec}$$

The horizontal distance between the box and the trip is

$$x = v_x t$$
$$= \left(100 \frac{\text{ft}}{\text{min}}\right)\left(12 \frac{\text{in}}{\text{ft}}\right)\left(\frac{1 \text{ min}}{60 \text{ sec}}\right)(0.70 \text{ sec})$$
$$= 14 \text{ in}$$

The answer is (A).

Why Other Options Are Wrong

(B) This incorrect solution is the time it takes for the part to drop all the way to the floor instead of into the box.

(C) This incorrect solution results when the conversions are neglected.

(D) This incorrect solution results when the function of velocity, instead of the initial velocity, is used.

SOLUTION 12

The angular velocity of the mechanism at 10 sec is

$$\omega_{10 \text{ sec}} = \alpha t$$
$$= \left(50 \ \frac{\text{rad}}{\text{sec}^2}\right)(10 \text{ sec})$$
$$= 500 \text{ rad/sec}$$

The circumference of the bolt circle is

$$C = \pi d$$
$$= \pi(12 \text{ in})$$
$$= 37.70 \text{ in}$$

The velocity of the bolt at 10 sec is

$$\text{v}_{10 \text{ sec}} = C\omega_{10 \text{ sec}}$$
$$= \left(37.70 \ \frac{\text{in}}{\text{rev}}\right)\left(500 \ \frac{\text{rad}}{\text{sec}}\right)\left(\frac{1 \text{ rev}}{2\pi \text{ rad}}\right)$$
$$\times \left(3600 \ \frac{\text{sec}}{\text{hr}}\right)\left(\frac{1 \text{ mi}}{5280 \text{ ft}}\right)\left(\frac{1 \text{ ft}}{12 \text{ in}}\right)$$
$$= 170.5 \text{ mph} \quad (170 \text{ mph})$$

The answer is (C).

Why Other Options Are Wrong

(A) This incorrect solution results when the diameter of the bolt circle, instead of the circumference, is used.

(B) This incorrect solution results when the radius, instead of the diameter, is used to calculate the circumference of the circle.

(D) This incorrect solution results when the angular velocity is not converted from radians to revolutions.

SOLUTION 13

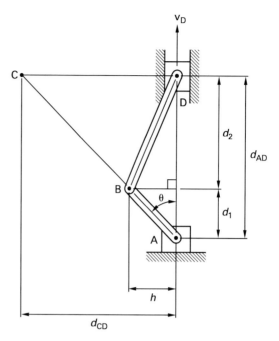

At any given instant, arm BD seems to rotate about an instantaneous center, point C, but with a different angular velocity from the crank. Point C lies at the intersection of lines perpendicular to the velocity vectors of points B and D; that is, at the intersection of line AB and the horizontal line through point D.

The velocity of point D at any instant can be calculated from the instantaneous angular velocity of the arm and the distance from point C.

$$\text{v}_\text{D} = d_\text{CD}\omega_\text{arm}$$

The angular velocity of the crank at 10 sec is

$$\omega_\text{crank} = \alpha t = \left(1.75 \ \frac{\text{rad}}{\text{sec}^2}\right)(10 \text{ sec})$$
$$= 17.5 \text{ rad/sec}$$

The angle of the crank at 10 sec is calculated from the decimal portion of the number of revolutions.

$$\theta_\text{rev} = \theta_0 + \omega_0 t + \frac{1}{2}\alpha_0 t^2$$
$$= 0 \text{ rev} + \left(0 \ \frac{\text{rad}}{\text{sec}}\right)(10 \text{ sec})$$
$$+ \left(\frac{1}{2}\right)\left(1.75 \ \frac{\text{rad}}{\text{sec}^2}\right)(10 \text{ sec})^2$$
$$= 87.5 \text{ rad}$$
$$N = \frac{\theta}{2\pi} = (87.5 \text{ rad})\left(\frac{1 \text{ rev}}{2\pi}\right)$$
$$= 13.93 \text{ rev}$$

0.93 rev is equivalent to -0.07 rev. The angle of the crank in relation to the vertical is

$$\theta_{10\,\text{sec}} = N_{\text{rev}}\left(\frac{360°}{\text{rev}}\right)$$
$$= (-0.07 \text{ rev})\left(\frac{360°}{\text{rev}}\right)$$
$$= -25.2°$$

With the angle of the crank at 10 sec known, and the lengths of the crank and the arm, the distance between points C and D can be calculated.

$$\cos\theta = \frac{d_1}{d_{AB}}$$
$$d_1 = d_{AB}\cos\alpha$$
$$= (6 \text{ in})\cos 25.2°$$
$$= 5.43 \text{ in}$$

$$\sin\theta = \frac{h}{d_{AB}}$$
$$h = d_{AB}\sin\alpha$$
$$= (6 \text{ in})\sin 25.2°$$
$$= 2.55 \text{ in}$$

$$h^2 + d_2^2 = d_{BD}^2$$
$$d_2 = \sqrt{d_{BD}^2 - h^2}$$
$$= \sqrt{(10 \text{ in})^2 - (2.55 \text{ in})^2}$$
$$= 9.67 \text{ in}$$

$$d_{AD} = d_1 + d_2$$
$$= 5.43 \text{ in} + 9.67 \text{ in}$$
$$= 15.10 \text{ in}$$

$$\tan\theta = \frac{d_{CD}}{d_{AD}}$$
$$d_{CD} = d_{AD}\tan\theta$$
$$= (15.10 \text{ in})\tan 25.2°$$
$$= 7.11 \text{ in}$$

The distance between points A and C is found from the equation

$$d_{AC}^2 = d_{AD}^2 + d_{CD}^2$$
$$d_{AC} = \sqrt{d_{AD}^2 + d_{CD}^2}$$
$$= \sqrt{(15.10 \text{ in})^2 + (7.11 \text{ in})^2}$$
$$= 16.69 \text{ in}$$

The distance between points B and C is

$$d_{BC} = d_{AC} - d_{AB}$$
$$= 16.69 \text{ in} - 6 \text{ in}$$
$$= 10.69 \text{ in}$$

The instantaneous angular velocity of arm BD around point C is found from the equation

$$\frac{\omega_{\text{arm}}}{\omega_{\text{crank}}} = \frac{d_{AB}}{d_{BC}}$$
$$\omega_{\text{arm}} = \frac{d_{AB}\omega_{\text{crank}}}{d_{BC}}$$
$$= \frac{(6 \text{ in})\left(17.5\,\dfrac{\text{rad}}{\text{sec}}\right)}{10.69 \text{ in}}$$
$$= 9.82 \text{ rad/sec}$$

The velocity of point D at 10 sec is

$$v_D = d_{CD}\omega_{\text{arm}}$$
$$= \left(\frac{7.11 \text{ in}}{12\,\dfrac{\text{in}}{\text{ft}}}\right)\left(9.82\,\dfrac{\text{rad}}{\text{sec}}\right)$$
$$= 5.82 \text{ ft/sec} \quad (5.8 \text{ ft/sec})$$

The answer is (C).

Why Other Options Are Wrong

(A) This incorrect solution results when the angular acceleration, instead of the angular velocity, is used.

(B) This incorrect solution results when d_{CD} is treated as a diameter through point C instead of a radius.

(D) This incorrect solution results when the number of turns is mislabeled as radians instead of revolutions.

SOLUTION 14

The effective throat size is

$$t_e = \frac{y}{\sqrt{2}} = \frac{0.375 \text{ in}}{\sqrt{2}}$$
$$= 0.2651 \text{ in}$$

The shear stress in the fillet weld is

$$\tau = \frac{F}{bt_e} = \frac{22{,}500 \text{ lbf}}{(10 \text{ in})(0.2651 \text{ in})}$$
$$= 8487 \text{ lbf/in}^2 \quad (8500 \text{ psi})$$

The answer is (D).

Why Other Options Are Wrong

(A) This incorrect solution results when the combined thickness of the two plates, instead of the effective throat size, is used.

(B) This incorrect solution results when the thickness of the weld, instead of the effective throat size, is used.

(C) This incorrect solution results when the thickness of one plate, instead of the effective throat size, is used.

SOLUTION 15

The angular velocity of the system is

$$\omega = \frac{v_{max}}{A}$$

$$= \frac{\left(61.5 \ \frac{ft}{min}\right)\left(\frac{1 \ min}{60 \ sec}\right)\left(12 \ \frac{in}{ft}\right)}{3 \ in}$$

$$= 4.1 \ rad/sec$$

The distance to the spring can be calculated from the formula

$$\omega = \sqrt{\frac{\left(\frac{m}{g_c}\right)gL + kd^2}{\left(\frac{m}{g_c}\right)L^2}}$$

$$= \sqrt{\frac{mgL + kg_c d^2}{mL^2}}$$

$$4.1 \ \frac{rad}{sec} = \sqrt{\frac{\begin{pmatrix}(5 \ lbm)\left(32.2 \ \frac{ft}{sec^2}\right) \\ \times \left(12 \ \frac{in}{ft}\right)(2 \ ft)\left(12 \ \frac{in}{ft}\right) \\ + \left(10 \ \frac{lbf}{in}\right)\left(32.2 \ \frac{ft\text{-}lbm}{lbf\text{-}sec^2}\right) \\ \times \left(12 \ \frac{in}{ft}\right)d^2 \end{pmatrix}}{(5 \ lbm)\left((2 \ ft)\left(12 \ \frac{in}{ft}\right)\right)^2}}$$

$$d = 0.72746 \ in \quad (0.73 \ in)$$

The answer is (B).

Why Other Options Are Wrong

(A) This incorrect solution results from neglecting all conversion factors.

(C) This incorrect solution results when the formula for a horizontal constrained compound pendulum is used instead of for a vertical one.

(D) This incorrect solution results when m is used instead of m/g_c throughout the formula for angular velocity.

SOLUTION 16

The maximum allowable stress for SA-515 65 plate at 500°F can be found in a table of maximum allowable stress, and is

$$S = 18{,}600 \ psi$$

The joint efficiency for a fully examined single butt weld with an integral backing strip can be found in a table of joint efficiencies, and is

$$E = 0.90$$

Check the validity of the pressure head design.

$$p \leq 0.665 SE$$

$$175 \ \frac{lbf}{in^2} \leq (0.665)\left(18{,}600 \ \frac{lbf}{in^2}\right)(0.90)$$

$$175 \ \frac{lbf}{in^2} \leq 11{,}132 \ \frac{lbf}{in^2}$$

The pressure design is valid. The thickness required is calculated from the equation

$$t = \frac{pr}{2SE - 0.2p}$$

$$= \frac{\left(175 \ \frac{lbf}{in^2}\right)\left(\frac{50 \ in}{2}\right)}{(2)\left(18{,}600 \ \frac{lbf}{in^2}\right)(0.9) - (0.2)\left(175 \ \frac{lbf}{in^2}\right)}$$

$$= 0.131 \ in$$

Check the validity of the thickness.

$$t \leq 0.356 r$$
$$0.131 \ in \leq (0.356)(25 \ in)$$
$$0.131 \ in \leq 8.9 \ in$$

The thickness is valid. The smallest standard plate thickness greater than 0.131 in is 0.188 in.

The answer is (B).

Why Other Options Are Wrong

(A) This incorrect solution results when the efficiency of the welded joint is taken as 1.00 for a double weld butt joint.

(C) This incorrect solution results when the diameter of the head, instead of the radius, is used.

(D) This incorrect solution results when the formula for the limitation is used to calculate the wall thickness of the head.

SOLUTION 17

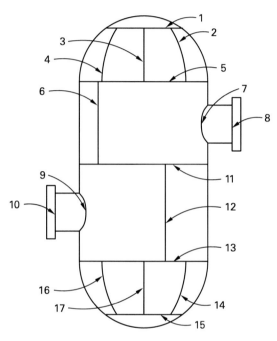

Category A welds include longitudinal joints in the main shell and diameter transitions. Welds 1, 2, 3, 4, 5, 6, 12, 13, 14, 15, 16, and 17 fit this description, for a total of 12 Category A welds.

Category B welds include circumferential welded joints in the main shell. Weld 11 fits the description for Category B welds.

Category C welds include welded joints connecting flanges. Welds 8 and 10 fit the description for Category C welds.

Category D welds include chambers or nozzles to the main shell. Welds 7 and 9 fit the description for Category D welds.

The answer is (B).

Why Other Options Are Wrong

(A) This incorrect solution results when only welds 2, 3, 4, 6, 12, 14, 16, and 17 are classified as Category A. Although welds 1, 5, 13, and 15 are circumferential welds in the main shell, they are also diameter transitions, which places them in Category A.

(C) This incorrect solution results when weld 11 is counted as a Category A weld. Weld 11 is a circumferential welded joint in the main shell and falls into Category B.

(D) This incorrect solution results when all welds are classified in Category A. Weld 11 is a circumferential welded joint in the main shell and falls into Category B. Welds 8 and 10 are welded joints connecting flanges, placing them into Category C. Welds 7 and 9 are connecting a chamber or nozzle to the main shell, classifying them as Category D.

SOLUTION 18

When two dissimilar materials are coupled, they form an electrolytic cell containing a cathode and an anode. The more electronegative material becomes the anode, which eventually dissolves and is considered the sacrificial material. Items I, II, and III reflect this information correctly.

The answer is (C).

Why Other Options Are Wrong

(A) This option leaves out item II, which is also true. The electronegative material will dissolve.

(B) This option includes item IV, which is incorrect. The anode is the material sacrificed in an electrolyte cell.

(D) This option includes item IV which is incorrect. The anode is the material sacrificed in an electrolytic cell.

SOLUTION 19

The rate of corrosion loss is found with the formula

$$R = \frac{87{,}600\,m}{\rho A t}$$

$$= \frac{\left(87{,}600\,\dfrac{\mu\text{m}\cdot\text{h}\cdot\text{g}}{\text{yr}\cdot\text{cm}\cdot\text{mg}}\right)(1.5\text{ g})\left(1000\,\dfrac{\text{mg}}{\text{g}}\right)}{\left(8602\,\dfrac{\text{kg}}{\text{m}^3}\right)\left(\dfrac{1\text{ m}}{100\text{ cm}}\right)^3} \\ \times\left(1000\,\dfrac{\text{g}}{\text{kg}}\right)(5\text{ cm}^2)(72\text{ h})$$

$$= 42{,}432\ \mu\text{m/yr}\quad(42{,}400\ \mu\text{m/yr})$$

The answer is (D).

Why Other Options Are Wrong

(A) This incorrect solution results when conversion factors are neglected.

(B) This incorrect solution results when the constant 87,600 is neglected.

(C) This incorrect solution results when density is not converted.

2 Depth Problems

PRINCIPLES

PROBLEM 1

A component of a machine consists of a 12 ft 6 in steel span of 6 in square tubing. A 10 ft 6 in linear slide bearing centered in the span supports an 800 lbf load. The maximum deflection of the span that will allow the bearing to move freely is 0.1 in. The beam is pinned at both ends, and a 2.6 factor of safety is required.

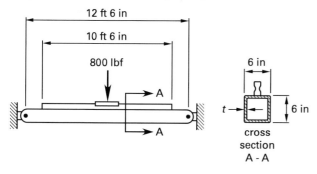

Assuming the weight of the steel can be neglected, what standard tubing wall thickness will satisfy the conditions?

(A) 0.188 in
(B) 0.250 in
(C) 0.375 in
(D) 0.500 in

Hint: The steel span is a simply supported beam pinned at both ends. The maximum deflection will be in the center of the span.

PROBLEM 2

A piece of solid square steel stock 4 ft long is cantilevered horizontally at the end of a slide as a stop to protect expensive measurement equipment from damage. One end is fixed and the other end is free. Young's modulus for the steel is 29×10^6 psi. Disregard bending and eccentric stresses. If the factor of safety against buckling is 2.6, what is the smallest hot-rolled stock size that should be used to absorb an axial force of 10,000 lbf without buckling?

(A) 1 in
(B) 1⅜ in
(C) 1¾ in
(D) 1⅞ in

Hint: A long, thin member in compression can be treated as a column. Use Euler's column equations.

PROBLEM 3

Which of the following statements are true about 1020 steel?

  I. 1020 is a low-carbon steel.
 II. 1020 contains 10% chromium.
III. 1020 contains 0.20% carbon.
 IV. 1020 contains 0.20% molybdenum.

(A) I and III only
(B) II and III only
(C) II and IV only
(D) I, II, and III

Hint: Review the AISI-SAE four-digit designations for typical steels and alloys.

PROBLEM 4

An 8 ft rod with a diameter of 0.5 in is used as a tie rod in a temperature-controlled environment. The tie rod is designed to fail when the tension reaches 2650 lbf and the rod is elongated by 0.1250 in. What material is best suited for this purpose?

(A) aluminum alloy
(B) cast iron
(C) ductile cast iron
(D) stainless steel

Hint: Calculate the strain on the tie rod to determine the required modulus of elasticity.

PROBLEM 5

Material samples are tensile-tested to predict future performance in torsion. Before testing, the samples have an average diameter of 0.125 in and a length of 3 in. After a tensile pull force of 1000 lbf during testing, the samples have an average diameter of 0.1248 in and a length of 3.0153 in. The material will be used as a 4 ft long, 1 in diameter shaft subject to a torque of 10,000 in-lbf. The twist that can be predicted for the shaft in use is most nearly

(A) $3.8°$
(B) $4.6°$
(C) $46°$
(D) $70°$

Hint: Use the dimensions from the tensile test to calculate the modulus of elasticity and Poisson's ratio.

PROBLEM 6

A polyethylene polymer chain contains 362 carbon atoms. Its degree of polymerization is most nearly

(A) 181
(B) 182
(C) 361
(D) 400

Hint: A polyethylene chain is a series of CH_2CH_2 chains terminated at both ends by OH− free radicals. Write out the molecular structure of polyethylene.

PROBLEM 7

Which of the following are examples of nondestructive testing?

I. eddy current testing
II. ultrasound imaging
III. scleroscopic hardness test
IV. Charpy test
V. Brinell hardness test

(A) I and II only
(B) IV and V only
(C) I, II, and III
(D) III, IV, and V

Hint: Review the definition of each test.

PROBLEM 8

Which of the following statements are true?

I. Brittle materials fail by sudden fracturing, not by yielding.
II. A ductile material should be designed for use with its maximum stress greater than its yield strength.
III. Brittle materials generally have much higher compressive strengths than tensile strengths.
IV. Resilient materials are able to absorb and release strain energy without experiencing permanent deformation.

(A) I and III only
(B) II and IV only
(C) I, II, and III
(D) I, III, and IV

Hint: Review the definitions of brittle materials, ductile materials, and resilient materials.

PROBLEM 9

A weldment of 1.5 in square steel stock consists of two 48 in pieces welded perpendicularly to a third piece 18 in long. The first weld occurs at 6 in, and the second occurs at 18 in. The unsupported weight of the steel causes the weldment to twist slightly.

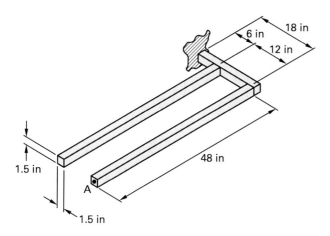

Neglecting any deflection of individual components, how much does point A, the end of the second welded piece, drop due to torsion from the weight of the weldment?

(A) 0.078 in

(B) 0.10 in

(C) 0.13 in

(D) 0.16 in

Hint: Calculate the end position of each welded piece as affected by torsion.

PROBLEM 10

To prevent casual adjustment, a fastener is designed to accept only a special elliptical torque driver. A 6 in long stainless steel elliptical torque driver, as shown, is designed to preload the fastener to 250 in-lbf. The stainless steel's yield strength in tension is 30,000 psi. The only device available to measure the torque during initial testing of the fastener is a protractor marked in 0.1° increments.

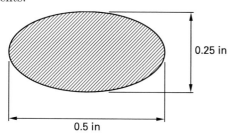

Which of the following statements is correct?

(A) The driver will not twist enough to register a measurement on the protractor.

(B) The protractor will read 0.4°.

(C) The protractor will read 6.6°.

(D) The driver will fail in shear before reaching the preload torque.

Hint: To compute the degree of twist and maximum shear in acoustic shapes, special torsion constants are needed.

PROBLEM 11

A small folding workbench is fixed to a wall and additionally supported by a guy wire to the left of a hinge. The right side rests on a wall support. The hinge is located 10 in from the wall. A 50 lbf weight is placed 4 in from the wall support.

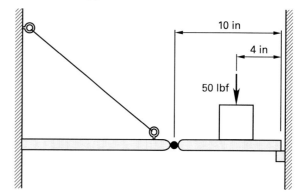

The reaction force at the hinge is most nearly

(A) 0 lbf

(B) 20 lbf

(C) 30 lbf

(D) indeterminate

Hint: The moment about a hinge is zero.

PROBLEM 12

A beam with a mass of 15 slugs, as shown, rolls on oiled steel tracks with a coefficient of friction of 0.1. The beam is moved back and forth via a rack and a pinion with a diameter of 1.25 in. A wheel is mounted on the pinion so that the applied force needed to move the beam is no more than 5 lbf.

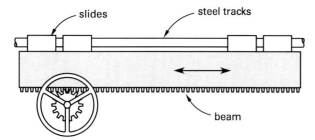

The minimum diameter of the wheel is most nearly

(A) 1.0 in

(B) 6.0 in

(C) 12 in

(D) 120 in

Hint: The ratio of a loadbearing force to an applied force is mechanical advantage.

PROBLEM 13

The simply supported machine structure shown supports a 500 lbf control box, a 15 lbf operator boom, and a welding curtain weighing 10 lbf/ft.

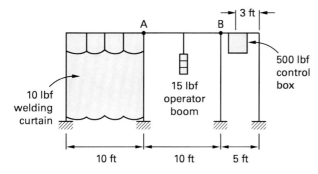

What is the moment at point B?

(A) −580 ft-lbf

(B) −170 ft-lbf

(C) −130 ft-lbf

(D) indeterminate

Hint: The machine support is a continuous beam. The problem can be solved using the three-moment method.

PROBLEM 14

Weights A and C are both accelerating downward at half of the acceleration of gravity. Weight B is supported by a block pulley.

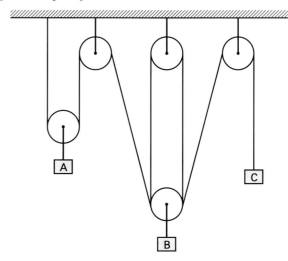

If positive numbers indicate upward acceleration, the acceleration of weight B is most nearly

(A) -12 ft/sec^2

(B) 12 ft/sec^2

(C) 24 ft/sec^2

(D) 32 ft/sec^2

Hint: The block-and-pulley arrangement is a dependent system. The relationship between the accelerations is the same as the relationship between the positions.

PROBLEM 15

A timing belt runs around an 18 in diameter pulley and has 60° of contact. The belt runs at 30 rpm. The belt's mass is 5 lbm/ft. The tight-side tension is 225 lbf, and the coefficient of friction is 0.37. The slack-side tension is most nearly

(A) 28 lbf

(B) 150 lbf

(C) 160 lbf

(D) 1600 lbf

Hint: Compute the tangential velocity first.

PROBLEM 16

A 3 in diameter disk flywheel with a mass of 0.5 lbm accelerates uniformly up from 0 rpm to 72 rpm in 5 sec. The work required to perform this is most nearly

(A) 0.041 in-lbf

(B) 0.17 in-lbf

(C) 16 in-lbf

(D) 45 in-lbf

Hint: Compute the work done by a torque of constant magnitude.

PROBLEM 17

A 3 oz ball is stationary on a track. A 5 oz ball is on the end of a 3 in pendulum arm as shown. The pendulum is powered by a spring compressed by 1 in. The larger ball strikes the smaller ball. After impact, the pendulum continues to move at 1.5 ft/sec, and the smaller ball moves along the track at 4 ft/sec. The spring constant needed to make this happen is most nearly

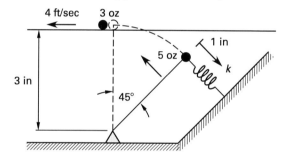

(A) 9.7 lbf/ft

(B) 28 lbf/ft

(C) 44 lbf/ft

(D) 450 lbf/ft

Hint: Use the conservation of momentum equation to calculate the initial velocity of the striking ball at impact.

PROBLEM 18

A 357 Magnum bullet (diameter of 0.357 in) weighing 80 grains and traveling 1800 ft/sec strikes a pendulum target. (7000 grains equals 1 lbm.)

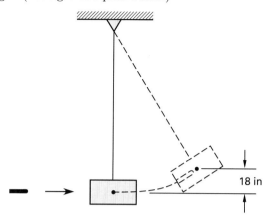

The minimum pendulum mass needed to keep the recoil height under 18 in is most nearly

(A) 0.20 lbm

(B) 0.59 lbm

(C) 2.1 lbm

(D) 32 lbm

Hint: When the bullet strikes the pendulum, momentum is conserved.

PROBLEM 19

A material flows through an extruder at a mass flow rate of 400 lbm/hr as the extruder operates at 40 rpm. To start a new process, the extruder ramps up from 0 rpm to 10 rpm at 0.25 rad/sec², then from 10 rpm to 15 rpm at 0.125 rad/sec², and finally from 15 rpm to 45 rpm at 0.375 rad/sec². The amount of material used in the ramp-up process is most nearly

(A) 0.36 kg

(B) 0.41 kg

(C) 0.84 kg

(D) 16 kg

Hint: Compute the number of revolutions for each ramp stage.

PROBLEM 20

As a filament is processed, its tension must be measured. To do this, one of the pulleys in the process is mounted on a load cell. A load cell has a collar to prevent damage when excessive tension overloads the pulley. The collar is expensive to replace, and in the current application the load cell is frequently vulnerable to loads that exceed the load rating. The pulley is inexpensive, so a sacrificial dowel of steel is positioned next to it. In the event of excessive tension, only the pulley and the dowel will be damaged, not the expensive collar and load cell. The rating of the load cell is 50 lbf/deg, and the maximum allowable deflection is 10°. The steel dowel must not deflect more than 0.01 in at the point of contact with the pulley in order to offer protection.

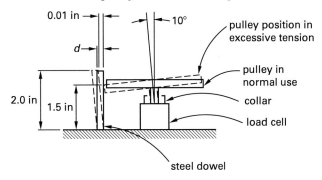

What minimum standard diameter is necessary for the sacrificial steel dowel to protect the load cell from forces up to 1000 lbf?

(A) 0.13 in

(B) 0.25 in

(C) 0.50 in

(D) 0.75 in

Hint: Treat the dowel as a cantilever beam to calculate diameter based on the allowed maximum deflection.

PROBLEM 21

A pair of 2 in by 10 in wood beams spanning 12 ft currently supports a 1000 lbf mezzanine. A new machine apparatus is affixed to the center of the double beam, adding 2500 lbf. Neglect the weight and strength of the wood.

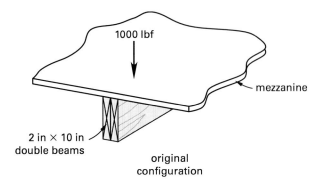

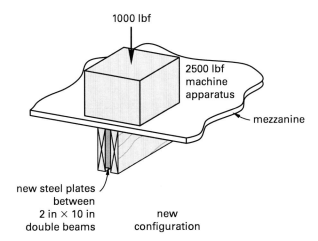

If the allowable stress is 24 ksi, how many 0.25 in by 9 in by 12 ft steel plates should be added between the wooden components to support the total weight?

(A) 1

(B) 2

(C) 4

(D) 7

Hint: Use the section modulus to determine the overall steel thickness required for the allowable stress.

PROBLEM 22

A simply supported 6 in by 8 in wood beam spanning 10 ft is supporting a distributed load. The beam is damaged 4 ft from the end, causing 2 in of material to be removed.

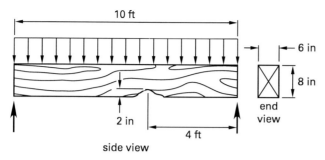

not to scale

If the allowable bending stress is 1200 psi, the percentage reduction in the load capacity at the damage is most nearly

(A) 25%
(B) 41%
(C) 44%
(D) 56%

Hint: Use moment and shear calculations to determine the distributed loads at the center of the span and at the damaged section.

PROBLEM 23

A timber cantilevered platform supporting a plastic compound hopper extends over an extruder. The cantilevered platform consists of 2×12 joists 16 in on center extending 80 in. The timber floor uses 1 in boards. The maximum weight of the plastic compound hopper and storage boxes distributes 500 lbf along each joist. The extruder operators are concerned for their safety while working under the old platform and want to make sure it is not overloaded. Using 35 lbf/ft^3 for the specific weight of the timber, the maximum bending stress on each joist is most nearly

(A) 120 psi
(B) 630 psi
(C) 690 psi
(D) 1300 psi

Hint: Determine the effect of each component of the platform and the weight of the plastic hopper on a single joist. Remember that the actual dimensions of wooden members differ from their nominal dimensions.

APPLICATIONS: MECHANICAL COMPONENTS

PROBLEM 24

What class of fit would be specified to achieve a constant pressure on a bearing?

(A) running and sliding fit
(B) locational fit
(C) force fit
(D) accurate fit

Hint: Classes of fits are arranged in three general groups.

PROBLEM 25

A machine twists a pair of copper wires at 5 rev/ft and runs at a speed of 300 ft/min. The twisted pair is wound by a mechanism supported by a deep-groove ball bearing. The bearing has a catalog-rated life of 3000 hr and a catalog-rated speed of 500 rpm. The load rating for the bearing and the application are the same. In machine hours, the bearing should be replaced every

(A) 15 hr
(B) 360 hr
(C) 1000 hr
(D) 3000 hr

Hint: Use the speed of the winding mechanism and the total number of bearing revolutions to determine the number of machine hours for each recommended service interval.

PROBLEM 26

A tapered roller bearing with a contact angle of 20° experiences combined radial and thrust forces as shown in the table. The inner race rotates, and shock and impact are negligible.

load phase	phase description	speed (rpm)	time (sec)	radial force (lbf)	thrust force (lbf)	radial factor X	thrust factor Y
1	load/advance	500	7	250	500	0.4	$0.4 \cot \alpha$
2	spin	5000	10	750	1000	0.4	$0.4 \cot \alpha$
3	retract/unload	250	3	−250	500	0.4	$0.4 \cot \alpha$

The dynamic equivalent radial load is most nearly

(A) 900 lbf
(B) 1300 lbf
(C) 1400 lbf
(D) 4700 lbf

Hint: Calculate the dynamic equivalent radial load for each phase and average.

PROBLEM 27

A ball bearing has a catalog load rating of 800 lbf and a catalog speed rating of 500 rpm for 3000 hr. The bearing application has a design radial load of 750 lbf and a design speed of 1500 rpm for 1000 hr. The relation constant for the bearing is 3. The reliability of the bearing for this application is most nearly

(A) 0.83
(B) 0.87
(C) 0.91
(D) 0.92

Hint: The reliability of a bearing is a logarithmic relationship based on the bearing's rating life and design life.

PROBLEM 28

For the journal bearing shown, the friction variable is 3.50.

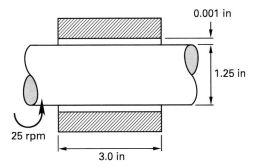

not to scale

The absolute viscosity required to support a 250 psi pressure at 25 rpm is most nearly

(A) 2.7×10^{-4} reyn
(B) 3.3×10^{-4} reyn
(C) 4.0×10^{-3} reyn
(D) 4.2×10^{-2} reyn

Hint: Find the coefficient of friction with the friction variable. Then use the frictional torque to solve for the shearing stress in the lubricant.

PROBLEM 29

A journal bearing has a shaft diameter of 1.5 in with 0.0015 in clearance. A 750 lbf load is applied as the shaft rotates at 30 rpm. The lubricant used has a specific heat of 0.42 Btu/lbm-°F and a density of 0.0311 lbm/in^3. The friction variable is 3.50. The lubricant flow rate needed to keep the increase in temperature under 30°F is most nearly

(A) 1.8×10^{-2} ft^3/sec
(B) 2.0×10^{-6} ft^3/sec
(C) 1.2×10^{-4} ft^3/sec
(D) 3.4×10^{-3} ft^3/sec

Hint: Calculate the generated heat to find the flow rate.

PROBLEM 30

For the gear train shown, the rotational speed, n, of gear 2 and the number of teeth, N, on each gear are as indicated. Gear 9 (the sun gear) is fixed.

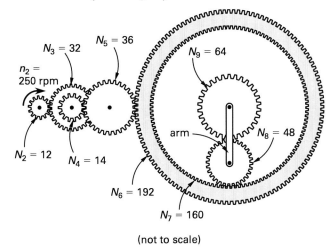

(not to scale)

The rotational speed of the arm around the sun gear is most nearly

(A) 4.9 rpm
(B) 5.3 rpm
(C) 6.8 rpm
(D) 11 rpm

Hint: For computational purposes, this system can be broken into a compound mesh gear train and a planetary gear arrangement.

PROBLEM 31

A 12 in diameter gear is spinning at 1250 rpm. 25 hp is applied to the 6 in diameter pinion.

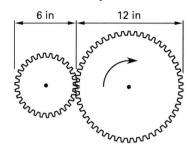

The torque applied to the pinion is most nearly

(A) 53 ft-lbf
(B) 83 ft-lbf
(C) 110 ft-lbf
(D) 210 ft-lbf

Hint: Use the speed of the pinion and the pitch velocity to calculate the transmitted load.

PROBLEM 32

In a set of straight bevel gears, the gear has 96 teeth and a pitch angle of 70°. How many teeth does the pinion have?

(A) 20
(B) 35
(C) 109
(D) 264

Hint: The tangent of the pitch angle is the ratio of the number of teeth of the gear and pinion.

PROBLEM 33

A bevel gear arrangement is shown. The speed of the pinion is raised to 300% of its former rate. The diameter of the pinion is reduced by 50%, and the transmitted tangential load, F_t, is raised to 400%.

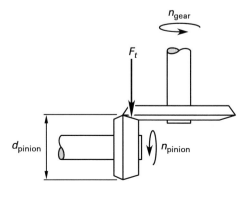

To maintain the system, the horsepower must be changed. The percentage of the original power required is most nearly

(A) 17%
(B) 270%
(C) 300%
(D) 600%

Hint: The solution can be found from the relationship of the variables without requiring the actual values.

PROBLEM 34

A pallet loader is a platform supported equally by four identical springs. After the top plate in the stack is removed, the loader raises the remaining plates to machine level in 0.250 in increments. The plates are 0.250 in thick and have a mass of 4.83 lbm. The springs are made of stainless steel wire (ASTM 313) with a wire diameter of 0.187 in. Each spring has 84 coils, with squared and ground ends. Centering rods prevent the springs from buckling.

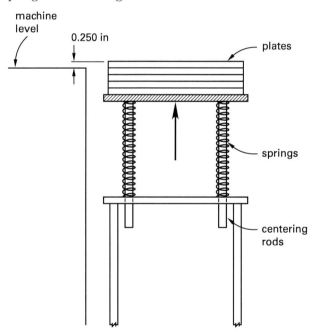

The maximum diameter of the centering rods is most nearly

(A) 0.375 in
(B) 0.750 in
(C) 1.25 in
(D) 2.00 in

Hint: Find the inside diameter of the spring.

PROBLEM 35

A worm gear has a lead of 4 in and a lead angle of 15°. The circular pitch of the mating gear is 1.25 in, and the center-to-center distance is 21.4 in. How many teeth must the mating gear have?

(A) 16
(B) 32
(C) 40
(D) 96

Hint: Begin by calculating the pitch diameters of each gear.

PROBLEM 36

A helical torsion spring has a spring index of 5. It takes 10 in-lbf of torque to twist the spring $3/4$ turn. The torsional spring constant is most nearly

(A) 1 in-lbf/rad
(B) 2 in-lbf/rad
(C) 5 in-lbf/rad
(D) 10 in-lbf/rad

Hint: The torsional spring constant is the torque needed to twist the spring.

PROBLEM 37

In an engineering competition, students try to launch water balloons over a 50 ft range using catapults and slingshots. The balloons must be launched horizontally from the roof of the engineering building, 60 ft above the ground. All springs used in the competition must have a spring constant of 3 lbf/in and be extended no more than 6 in. The weight of each projectile water balloon is 4 lbf. Under these conditions, what is the least number of springs in parallel that can send a balloon 50 ft?

(A) 3
(B) 10
(C) 17
(D) 25

Hint: Find the kinetic energy that must be imparted to the balloon.

PROBLEM 38

A helical compression spring has a wire diameter of 0.125 in and an outside diameter of 1.375 in. The spring is dynamically loaded. The cycle is forced between 2 lbf and 5 lbf. The maximum stress of the spring is most nearly

(A) 140 psi
(B) 4400 psi
(C) 6000 psi
(D) 8800 psi

Hint: Calculate the stress amplitude and the mean stress of the spring.

PROBLEM 39

In a cyclic operation, an A313 stainless steel wire helical spring has an outside diameter of 0.5 in and a wire diameter of 0.08 in with 30 active coils. Critical frequency is most nearly

(A) 60 Hz
(B) 100 Hz
(C) 200 Hz
(D) 400 Hz

Hint: Use the spring index and the weight of the active coils to find the critical frequency.

PROBLEM 40

Two plates 25 in wide and 0.0625 in thick are connected (along their 25 in edges) by a lap joint. The allowable stress of each 0.5 in diameter bolt is 16,250 psi. A force of 25,000 lbf is placed on the joint, and no more bolts are used than are necessary to bear this. The tensile stress on the plates is most nearly

(A) 1000 psi
(B) 8000 psi
(C) 19,000 psi
(D) 25,000 psi

Hint: Calculate the shear on each bolt and the number of bolts required to get the maximum tensile stress on the plate.

PROBLEM 41

A particular piece is available in four competing models, with specifications as shown.

model	I	II	III	IV
spindle diameter (in)	0.88	1.25	1.00	1.12
power rating (hp)	19	22	28	25

Spindle maximum shear stress is 2500 psi. The model with the fastest spindle speed is

(A) model I

(B) model II

(C) model III

(D) model IV

Hint: Use the shaft diameter and power to calculate the maximum rpm each unit achieves.

PROBLEM 42

A steel shaft has a diameter of 1.5 in and a length of 48 in and is supported at each end by self-aligning bearings. The weight of the shaft is 0.9 lbf/in. If the critical frequency needed to be raised at least 10 Hz higher than it is now, most nearly what would be the minimum diameter for the shaft?

(A) 1.6 in

(B) 1.7 in

(C) 1.9 in

(D) 2.1 in

Hint: Use a beam deflection formula to find the static deflection of the shaft.

PROBLEM 43

A 25 hp motor turns a shaft at 5000 rpm. The shaft is 15 in long, and it has an outside diameter of 2.5 in and an inside diameter of 1.5 in. The shear stress at the outer surface of the shaft is most nearly

(A) 15 psi

(B) 100 psi

(C) 120 psi

(D) 500 psi

Hint: Find the polar moment of inertia and the torque applied to the shaft by the motor.

PROBLEM 44

A steel stepped shaft is dimensioned as shown. The steel has a shear modulus of 11.5×10^6 psi. The equivalent torsional spring constant of the shaft is most nearly

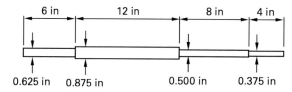

(A) 180 in-lbf

(B) 500 in-lbf

(C) 2900 in-lbf

(D) 98,000 in-lbf

Hint: Find the torsional spring constant for each step of the shaft.

PROBLEM 45

A water storage system consists of a primary tank, a secondary tank, and a 3 in diameter connecting pipe. Water is pumped through the pipe from the secondary to the primary tank. The pipe is straight, level, has an effective length of 32 ft, and has a friction factor of 0.03. Entrance and exit resistances are negligible, and flow velocity is 20 ft/sec. The head loss is most nearly

(A) 2.0 ft

(B) 19 ft

(C) 24 ft

(D) 77 ft

Hint: Calculate head loss using the Darcy equation.

APPLICATIONS: JOINTS AND FASTENERS

PROBLEM 46

A series of $\frac{1}{4}$-20 UNC threaded steel rods holds 12 in of rigid material together. Each rod is preloaded by precision tools in 0.1° increments to 1200 lbm. The number of degrees of turn required to properly preload the rod is most nearly

(A) 0.3°

(B) 5°

(C) 70°

(D) 100°

Hint: Find the change in length of the bolt, and then use the pitch to find the degrees of turn required.

PROBLEM 47

A 500 lbf load is applied as shown.

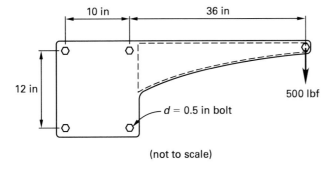

(not to scale)

Using an elastic analysis, the stress in the most critical bolt is most nearly

(A) 640 psi

(B) 3300 psi

(C) 3800 psi

(D) 3900 psi

Hint: The direction of the shear stress is the same as the rotation of the connection. Find the shear stresses that will account for this total stress.

PROBLEM 48

A single lap rivet joint is loaded in tension as shown. The sheet and rivet materials have the same properties: a tensile strength of 10,000 psi, a shear strength of 8500 psi, and a bearing strength of 20,000 psi.

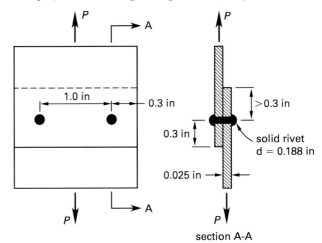

section A-A

The failure load is most nearly

(A) 180 lbf

(B) 190 lbf

(C) 400 lbf

(D) 470 lbf

Hint: The four most likely methods of failure are sheet bearing, rivet shear, sheet shear, and sheet tension.

PROBLEM 49

Four connection types are shown.

I.

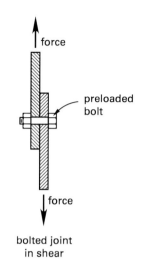

bolted joint
in shear

II.

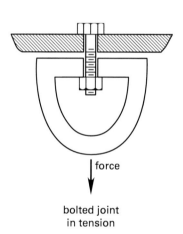

bolted joint
in tension

III.

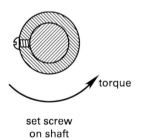

set screw
on shaft

IV.

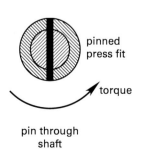

pin through
shaft

Which of these connections rely on friction for functionality?

(A) I and II only

(B) I and III only

(C) I, II, and III only

(D) I, II, III, and IV

Hint: Identify the action for each fastener.

PROBLEM 50

A machine frame is to be made of welded aluminum box beams cut to length and joined so that beam axes are perpendicular. The end of one beam mates with the face of another, and all four mating edges are completely welded with a fillet weld. Which of the following depicts a symbol for an acceptable weld under these conditions?

(A)

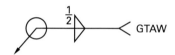

(B)

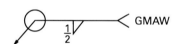

(C)

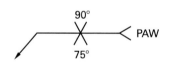

(D)

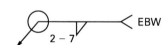

Hint: Review the meaning of each of the weld symbols.

APPLICATIONS: VIBRATION/DYNAMIC ANALYSIS

PROBLEM 51

A 30 in steel shaft is supported at each end by bearings. A 250 lbf gear is located 9 in from one end. The 1.375 in diameter shaft is specified to operate at no more than half the critical speed.

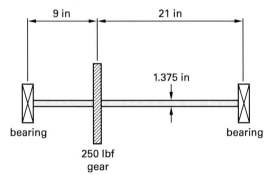

The maximum operating speed is most nearly

(A) 60 rad/sec

(B) 72 rad/sec

(C) 280 rad/sec

(D) 1400 rad/sec

Hint: The critical speed is the natural frequency of vibration for the shaft.

PROBLEM 52

A 5 lbm sheave operates between 50 rpm and 100 rpm. Belt tension is maintained by supporting the sheave on a spring-loaded hanger. The spring stiffness necessary to keep the resonance frequency of the sheave above the operating range is most nearly

(A) 0.43 lbf/ft

(B) 0.53 lbf/ft

(C) 17 lbf/ft

(D) 550 lbf/ft

Hint: At resonance, the forcing frequency equals the natural frequency.

PROBLEM 53

A 7500 lbm reel rotates at 70 rpm on center. At one end of the 90 in diameter reel, a 20 lbm reinforcement bar has been bolted on to strengthen a fractured rib. The natural rubber isolator has a damping factor of 0.05, and the spring constant of the structure is 30,000 lbf/in.

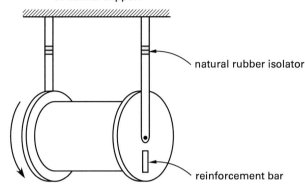

The maximum force transmitted by the reel onto the structure is most nearly

(A) 120 lbf

(B) 130 lbf

(C) 160 lbf

(D) 510 lbf

Hint: The force of the imbalance is created by the mass of the added bar, not the reel itself.

PROBLEM 54

A metal stamping machine drops a mass from a height of 24 in. The isolator on the machine recoils 6 in from the impact of the inelastic drop. 5000 lbf of force is necessary to punch a part. The minimum mass needed to punch a part is most nearly

(A) 19 lbm

(B) 160 lbm

(C) 630 lbm

(D) 3900 lbm

Hint: The mass is subjected to an inelastic drop. The acceleration of the mass can be measured in gravities.

PROBLEM 55

A binder is applied around a linear cable to hold components together as they are processed. The binder machine spins around the centerline of the processed cable and must be balanced for the machine to operate smoothly. The binder application assembly consists of the binder package and a die arrangement weighing 15 lbm and 10 lbm, respectively. These components are located 8 in apart opposite each other on a hollow shaft that rotates around the cable. The binder extends 4 in from the shaft, and the die extends 3 in from the shaft.

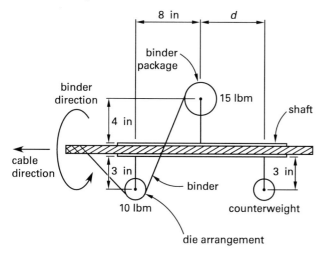

How far from the binder package should a counterbalance weight with an extension arm of 3 in be positioned?

(A) −4.0 in

(B) 4.0 in

(C) 8.0 in

(D) 16 in

Hint: Before calculating the counterbalance's position, calculate its weight.

APPLICATIONS: MATERIALS AND PROCESS

PROBLEM 56

A part is subjected to a 2500 psi stress alternating at 500 psi. A second orthogonal stress of 750 psi alternating at 1500 psi is added. The mean stress and alternating von Mises stress, respectively, are most nearly

(A) 1600 psi and 880 psi

(B) 1600 psi and 1000 psi

(C) 2200 psi and 1300 psi

(D) 3300 psi and 2000 psi

Hint: Use the distortion energy theory for biaxial loading.

PROBLEM 57

During heat treatment, which of the following are effects on tool steel of an extended exposure during the austenitizing temperature soak?

I. deterioration of structure

II. loss of hardness

III. carburization

IV. loss of magnetism

(A) III only

(B) I, II, and III

(C) I, II, and IV

(D) I, III, and IV

Hint: Review the definitions of structural deterioration, hardness, carburization, and magnetism.

PROBLEM 58

A part is subjected to a repeating pattern of 5 cycles at one stress level and then 10 cycles at another. The fatigue life is 10^7 cycles for the first stress level and 10^6 cycles for the second. The failure criterion is 0.90. The number of times the part can sustain the pattern is most nearly

(A) 0.86×10^5 repetitions

(B) 1.1×10^5 repetitions

(C) 1.6×10^5 repetitions

(D) 6.6×10^5 repetitions

Hint: Determine the relationship between the number of cycles at each stress level. The relationship can be inserted into Miner's rule (also known as the Palmgren-Minor cycle ratio summation formula) to compute the number of pattern cycles.

PROBLEM 59

Hot fluid is pumped through a 90% effective forced-convection heat exchanger. The hot fluid temperature drops 50°F across the heat exchanger, and the temperature difference between the hot and cold fluids entering the heat exchanger is 100°F. The thermal capacity rate of the hot fluid is 2.0 Btu/sec-°F, which is greater than the thermal capacity rate of the cold fluid. The specific heat of the cold fluid is 1.0 Btu/lbm-°F. The 95% efficient pump adds 5000 ft of pressure head. The required pump motor horsepower to pump the cold fluid is most nearly

(A) 9.6 hp

(B) 10 hp

(C) 11 hp

(D) 21 hp

Hint: Solve using the NTU method.

PROBLEM 60

A steam plant includes a 90% efficient boiler, a turbine, a condenser, and a feed pump. The plant consumes 5 tons of coal per day. The boiler exit temperature is 1000°F, the condenser exit temperature is 100°F, and the coal's higher heating value is 11,000 Btu/lbm. The feed pump requires 5% of the maximum theoretical power output. The theoretical net power output is most nearly

(A) 2.4×10^6 Btu/hr

(B) 2.7×10^6 Btu/hr

(C) 3.5×10^6 Btu/hr

(D) 3.9×10^6 Btu/hr

Hint: Determine the Carnot efficiency for this vapor power cycle problem.

PROBLEM 61

Which of the following statements are true regarding the statistical quality control chart shown?

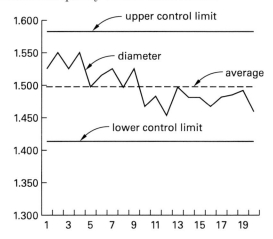

I. A suspicious event has occurred.

II. The data have shifted.

III. The data show a trend of decreasing value.

IV. The data are bimodal.

(A) I and IV only

(B) IV only

(C) I, II, and III

(D) II and III only

Hint: Review the definitions of a statistical process control chart.

PROBLEM 62

A cylindrical tube is composed of SAE 1020 steel and contains 10 lbm carbon dioxide at temperature of 70°F. The cylinder has a 4 in inner diameter, is 10 ft long, and is classified as a thick-wall cylinder. The cylinder's allowable stress is 5000 psi. Assume that carbon dioxide may be treated as an ideal gas. The minimum allowable thickness of the cylinder with both ends closed is most nearly

(A) 0.069 in

(B) 0.40 in

(C) 0.70 in

(D) 1.5 in

Hint: Use the ideal gas law to find the pressure.

PROBLEM 63

A plant's chief engineer must decide whether to buy or lease a portable HVAC system. The $10,000 system is expected to operate 3000 hr/yr and last 10 yr. If purchased, preventative maintenance will cost $0.25/hr of operation, the salvage value will be $1000, and corrective maintenance estimates include a repair cost of $50/hr, a mean time between failures of 3002 hr, and a mean time to failure of 3000 hr. If leased, a fee of $0.40/hr of operation covers all costs. Assume an interest rate of 7%, and use the year-end convention for all costs. At some point, the expected cost of buying will be the same as the expected cost of leasing. The time it will take to reach this point is most nearly

(A) 3.1 yr

(B) 3.9 yr

(C) 4.3 yr

(D) 4.5 yr

Hint: This is a break-even quantity analysis problem.

PROBLEM 64

A portable distillation plant with an acquisition cost of $3000 will have a salvage value of $1000 after five years. The plant is to be reconfigured upon installation at a cost of $1000. The special reconfiguration will provide an additional value of $1000 after five years. Routine maintenance will cost $400 per year. The effective annual interest rate is 5%. The average annual cost of ownership during the next five years is most nearly

(A) $560

(B) $760

(C) $960

(D) $17,000

Hint: Convert the present cost and future value of the plant to an equivalent uniform annual cost of ownership (EUAC).

PROBLEM 65

At the base of a liquid storage tank, a 1 ft^2 horizontal floor plate exerts force on a mechanical spring valve in proportion to the tank liquid level. The spring deflection is 3 in, the spring constant is 1000 lbf/in, and the liquid specific weight is 62.4 lbf/ft^3. The liquid level is most nearly

(A) 4.0 ft

(B) 5.3 ft

(C) 16 ft

(D) 48 ft

Hint: Equate spring force to fluid pressure force on a horizontal plane surface.

PROBLEM 66

It currently costs $120/hr to process a $0.65/lbm material. A cost-reduction project has identified a way to modify the process to use 10% less of the material, but at a higher cost of $210/hr. A feasibility study has identified a different usable material at $0.53/lbm, but that material is compatible only with the current process. A one-time qualification cost of $2 million is required to change either the process or the material. To remain competitive, the material supplier has offered a price reduction on the current material to $0.55/lbm upon payment of a $100,000 contract settlement fee; the price reduction is offered on the condition that the material continues to be used at the current rate. The process runs in a 24-hour/7-day-per-week plant environment with a machine efficiency of 75% and a material consumption rate of 2500 lbm/hr.

Total cost over 5 yr is the deciding factor. Which option should be selected?

(A) Option A: no changes

(B) Option B: new process, old material, old price

(C) Option C: old process, new material

(D) Option D: old process, old material, new price

Hint: Create a table itemizing the parameters. Then total the costs.

SOLUTION 1

The length of the steel span, converted from feet to inches, is

$$L = (12 \text{ ft})\left(12 \frac{\text{in}}{\text{ft}}\right) + 6 \text{ in}$$
$$= 150 \text{ in}$$

The load applied to the span is equivalent to

$$F = W(\text{FS}) = (800 \text{ lbf})(2.6)$$
$$= 2080 \text{ lbf}$$

The modulus of elasticity for steel is

$$E = 30 \times 10^6 \text{ psi}$$

Find the moment of inertia, using the formula for the maximum deflection of a simply supported beam pinned at each end with a center load.

$$y_{\max} = \frac{-FL^3}{48EI}$$

$$-0.1 \text{ in} = \frac{-(2080 \text{ lbf})(150 \text{ in})^3}{(48)\left(30 \times 10^6 \frac{\text{lbf}}{\text{in}^2}\right)I}$$

$$I = 48.75 \text{ in}^4$$

Use the moment of inertia to solve for wall thickness.

$$I = \frac{bh^3}{12}$$
$$= \frac{b_o h_o^3}{12} - \frac{b_i h_i^3}{12}$$

$$48.75 \text{ in}^4 = \frac{(6 \text{ in})(6 \text{ in})^3}{12} - \frac{(6 \text{ in} - 2t)(6 \text{ in} - 2t)^3}{12}$$
$$= \frac{(6 \text{ in})^4}{12} - \frac{(6 \text{ in} - 2t)^4}{12}$$
$$t = 0.42 \text{ in}$$

This answer must be rounded up to the next available tubing wall thickness, 0.500 in.

The answer is (D).

Why Other Options Are Wrong

(A) This incorrect solution results when the length is not raised to the third power in the deflection formula.

(B) This incorrect solution results when the length of the linear slide is used in place of the length of the steel span.

(C) This incorrect solution results when the factor of safety is applied to the wall thickness at the end of the solution, instead of being applied to the load on the bearing.

SOLUTION 2

The column is fixed at one end and free at the other. The end restraint coefficient, C, is 2.

Assume the stock behaves as a slender column. The moment of inertia may be found from the formula

$$(\text{FS})F = \frac{\pi^2 EI}{(CL)^2}$$

$$I = \frac{(CL)^2(\text{FS})F}{\pi^2 E}$$

$$= \frac{\left[(2)(4 \text{ ft})\left(12 \frac{\text{in}}{\text{ft}}\right)\right]^2 (2.6)(10{,}000 \text{ lbf})}{\pi^2 \left(29 \times 10^6 \frac{\text{lbf}}{\text{in}^2}\right)}$$

$$= 0.8372 \text{ in}^4$$

The thickness of the stock is found from the formula

$$I = \frac{bh^3}{12} = \frac{t^4}{12}$$
$$t = \sqrt[4]{12I}$$
$$= \sqrt[4]{(12)(0.8372 \text{ in}^4)}$$
$$= 1.78 \text{ in}$$

Hot-rolled square stock longer than 1 in is available in $\frac{1}{8}$ in increments. The standard size of stock is $1\frac{7}{8}$ in, or 1.875 in.

Check the assumption of a slender column. The radius of gyration is

$$k = \sqrt{\frac{I}{A}} = \sqrt{\frac{\frac{bh^3}{12}}{bh}}$$
$$= \sqrt{\frac{h^2}{12}}$$
$$= \sqrt{\frac{(1.875 \text{ in}^2)}{12}}$$
$$= 0.5413 \text{ in}$$

The slenderness ratio is

$$\frac{L}{k} = \frac{(4 \text{ ft})\left(12 \frac{\text{in}}{\text{ft}}\right)}{0.5413 \text{ in}}$$
$$= 88$$

The piece of stock has a slenderness ratio greater than 80 and can be treated as a slender column.

The answer is (D).

Why Other Options Are Wrong

(A) This incorrect solution results when the factor of safety is neglected.

(B) This solution results when the end conditions are neglected or incorrectly chosen as 1.

(C) This incorrect solution results when the nearest standard thickness is chosen even though it is slightly *less* thick than needed.

SOLUTION 3

According to AISI-SAI four-digit designations, carbon steels are designated as 10XX, 11XX, and 12XX. The last two digits indicate the percentage of carbon in hundredths of a percent. 1020 steel is thus a carbon steel with 0.20% carbon. Low-carbon steels contain less than 0.30% carbon. Statements I and III are true.

Chromium is added to steel to increase hardness, strength, and corrosion resistance, creating a steel alloy. Carbon steels do not contain chromium. 10% added chromium is found in heat- and corrosion-resistant steels such as stainless steel. Statement II is not true.

Molybdenum is not added to 1020 steel. Molybdenum is added to increase the high-temperature strength and hardness of a steel alloy. Statement IV is not true.

The answer is (A).

Why Other Options Are Wrong

(B) This option is wrong because 1020 steel is classified as a low-carbon steel, and because carbon steels do not contain chromium.

(C) This option is wrong because 1020 steel does not contain molybdenum or chromium.

(D) This option is wrong because carbon steels do not contain chromium.

SOLUTION 4

The original cross-sectional area of the tie rod is

$$A_0 = \pi\left(\frac{d}{2}\right)^2 = \pi\left(\frac{0.5 \text{ in}}{2}\right)^2$$
$$= 0.19635 \text{ in}^2$$

The cross-sectional area of the tie rod at failure is

$$A_f = \frac{A_0 L_0}{L_f} = \frac{(0.19635 \text{ in}^2)(8 \text{ ft})\left(12 \frac{\text{in}}{\text{ft}}\right)}{(8 \text{ ft})\left(12 \frac{\text{in}}{\text{ft}}\right) + 0.125 \text{ in}}$$
$$= 0.19609 \text{ in}^2$$

The true strain is

$$\epsilon = \ln \frac{A_0}{A_f}$$
$$= \ln \frac{0.19635 \text{ in}^2}{0.19609 \text{ in}^2}$$
$$= 0.001325$$

The applied stress is

$$\sigma = \frac{F}{A_0} = \frac{2650 \text{ lbf}}{0.19635 \text{ in}^2}$$
$$= 13,500 \text{ psi}$$

The needed modulus of the design material is

$$E = \frac{\sigma}{\epsilon}$$
$$= \frac{13,500 \frac{\text{lbf}}{\text{in}^2}}{0.001325}$$
$$= 10.3846 \times 10^6 \text{ lbf/in}^2 \quad (10.4 \times 10^6 \text{ psi})$$

From a table of moduli of elasticity, 10.4×10^6 psi is within the range of aluminum alloys, but is significantly below the ranges of cast iron, ductile cast iron, and stainless steel. The material best suited for this purpose is an aluminum alloy.

The answer is (A).

Why Other Options Are Wrong

(B) This incorrect solution results when the diameter, instead of the area, is used to determine the true strain.

(C) This incorrect solution results when the log, instead of the natural log, is used to calculate true strain.

(D) This incorrect solution results when the shear modulus, instead of the modulus of elasticity, is used to select a material.

SOLUTION 5

The average cross-sectional area before testing is

$$A_0 = \pi r^2 = \pi \left(\frac{d_0}{2}\right)^2$$
$$= \pi \left(\frac{0.125 \text{ in}}{2}\right)^2$$
$$= 0.01227 \text{ in}^2$$

The sample strain is

$$\epsilon = \frac{\Delta L}{L_0} = \frac{3.0153 \text{ in} - 3 \text{ in}}{3 \text{ in}}$$
$$= 0.0051$$

The stress applied to the sample is

$$\sigma = \frac{F}{A_0} = \frac{1000 \text{ lbf}}{0.01227 \text{ in}^2}$$
$$= 81,500 \text{ lbf/in}^2$$

The modulus of elasticity for the material is

$$E = \frac{\sigma}{\epsilon} = \frac{81,500 \dfrac{\text{lbf}}{\text{in}^2}}{0.0051}$$
$$= 15,980,000 \text{ lbf/in}^2$$

Poisson's ratio is

$$\nu = \frac{\epsilon_{\text{lateral}}}{\epsilon_{\text{axial}}} = \frac{\dfrac{\Delta d}{d_0}}{\dfrac{\delta}{L_0}}$$
$$= \frac{\dfrac{0.125 \text{ in} - 0.1248 \text{ in}}{0.125 \text{ in}}}{\dfrac{3.0153 \text{ in} - 3 \text{ in}}{3 \text{ in}}}$$
$$= 0.3137$$

The shear modulus is

$$G = \frac{E}{2(1+\nu)} = \frac{15,980,000 \dfrac{\text{lbf}}{\text{in}^2}}{(2)(1+0.3137)}$$
$$= 6,082,000 \text{ lbf/in}^2$$

The polar moment of inertia for the shaft is

$$J = \frac{\pi r_{\text{shaft}}^4}{2} = \frac{\pi \left(\dfrac{d_{\text{shaft}}}{2}\right)^4}{2}$$
$$= \frac{\pi \left(\dfrac{1 \text{ in}}{2}\right)^4}{2}$$
$$= 0.09817 \text{ in}^4$$

The predicted angle of twist for the shaft is

$$\gamma = \frac{TL}{JG}$$
$$= \left(\frac{(10,000 \text{ in-lbf})(4 \text{ ft})\left(12 \dfrac{\text{in}}{\text{ft}}\right)}{(0.09817 \text{ in}^4)\left(6,082,000 \dfrac{\text{lbf}}{\text{in}^2}\right)}\right)\left(\frac{360°}{2\pi}\right)$$
$$= 46.0616° \quad (46°)$$

The answer is (C).

Why Other Options Are Wrong

(A) This incorrect solution results when the length of the shaft is not converted from feet to inches.

(B) This incorrect solution results when the pulling force of the tensile test is used for the torque of the shaft.

(D) This solution results when Poisson's ratio is calculated incorrectly, using the new diameter and length instead of the change in diameter and length.

SOLUTION 6

The molecular structure of polyethylene is

$$\text{OH} - \underset{\underset{\text{H}}{|}}{\overset{\overset{\text{H}}{|}}{\text{C}}} - \underset{\underset{\text{H}}{|}}{\overset{\overset{\text{H}}{|}}{\text{C}}} - \cdots - \underset{\underset{\text{H}}{|}}{\overset{\overset{\text{H}}{|}}{\text{C}}} - \underset{\underset{\text{H}}{|}}{\overset{\overset{\text{H}}{|}}{\text{C}}} - \text{OH}$$

The chain contains 362 carbon atoms, and one oxygen atom at each end for a total of two. There are two hydrogen atoms for each carbon atom, plus one more at each end, so

$$N_{\text{H}} = 2N_{\text{C}} + 2$$
$$= (2)(362) + 2$$
$$= 726$$

The molecular weights of the elements carbon, hydrogen, and oxygen are equal to their atomic weights.

$$MW_C = A_C = 12.01115 \text{ lbm/lbmol}$$
$$MW_H = A_H = 1.00797 \text{ lbm/lbmol}$$
$$MW_O = A_O = 15.9994 \text{ lbm/lbmol}$$

The molecular weight of the polymer is

$$\begin{aligned} MW_{polymer} &= N_C(MW_C) + N_H(MW_H) + N_O(MW_O) \\ &= (362)\left(12.01115 \, \frac{\text{lbm}}{\text{lbmol}}\right) \\ &\quad + (724)\left(1.00797 \, \frac{\text{lbm}}{\text{lbmol}}\right) \\ &\quad + (2)\left(15.994 \, \frac{\text{lbm}}{\text{lbmol}}\right) \\ &= 5109.8 \text{ lbm/lbmol} \end{aligned}$$

The molecular weight of the mer, or repeating unit, is

$$\begin{aligned} MW_{mer} &= N_C(MW_C) + N_H(MW_H) \\ &= (2)\left(12.01115 \, \frac{\text{lbm}}{\text{lbmol}}\right) \\ &\quad + (4)\left(1.00797 \, \frac{\text{lbm}}{\text{lbmol}}\right) \\ &= 28.054 \text{ lbm/lbmol} \end{aligned}$$

The degree of polymerization is

$$\begin{aligned} DP &= \frac{MW_{polymer}}{MW_{mer}} \\ &= \frac{5109.8 \, \frac{\text{lbm}}{\text{lbmol}}}{28.054 \, \frac{\text{lbm}}{\text{lbmol}}} \\ &= 182.14 \quad (182) \end{aligned}$$

The answer is (B).

Why Other Options Are Wrong

(A) This incorrect solution results when the OH$^-$ terminators are neglected and only the mer is considered.

(C) This incorrect solution results when only half the mer is considered. The ethylene mer is CH$_2$CH$_2$.

(D) This incorrect solution results when the degree of polymerization is calculated for a polymer with only one mer and then multiplied by the number of mers.

SOLUTION 7

A test is destructive if it damages or mars the material being tested. Nondestructive testing is performed when destructive methods are not feasible. Such cases include the testing of one-of-a-kind items, manufactured products, and items involving expensive material or fabrication costs.

Eddy current testing uses alternating current from a test coil to induce eddy currents in electrically conductive items. Changes in current flow are used to detect flaws and other nonconforming material properties such as cracks, weld defects, voids, hardness, structure, and porosity. Option I is correct, as there is no damage to the test specimen.

Ultrasound imaging testing uses transmitted vibrations that are reflected by the object. Defects cause these vibration waves to scatter, making this technique appropriate for detecting inclusions, cracks, porosity, and changes in material structure. Option II is correct, as there is no damage to the test specimen.

Scleroscopic hardness testing is a rebound test. A controlled mass is dropped from a known height onto the workpiece. The height of the rebound correlates to hardness scales. Option III is correct, as there is no damage to the test specimen.

The Charpy test measures toughness on sample specimens. A 45° notch is cut into a standardized beam, which is centered on supports with the notch facing down. A falling striker hits the center of the specimen. The test is performed from different heights until a specimen fractures. Option IV is incorrect, as this test requires a standardized sample, an irreversible change (the notch) to the sample, and that the sample be tested to failure.

The Brinell hardness test presses a hardened steel ball into the surface of the workpiece. The Brinell hardness number is the quotient of the applied load and the area of the depression created by the penetrator. Option V is incorrect, as this test leaves an indentation on the workpiece.

The answer is (C).

Why Other Options Are Wrong

(A) This answer is incorrect because the scleroscopic hardness test is nondestructive.

(B) This answer is incorrect because the Charpy and Brinell hardness tests are destructive.

(D) This answer is incorrect because the Charpy and Brinell hardness tests are destructive.

SOLUTION 8

Statements I, III, and IV are correct. A ductile material should always be designed for use with its maximum stress *lower than* its yield strength.

The answer is (D).

Why Other Options Are Wrong

(A) This answer is wrong because it excludes statement IV, which is true. Resilient materials can absorb and release strain energy without permanent deformation.

(B) This answer is wrong because it includes statement II, which is false, and also because brittle materials do fail by sudden fracturing.

(C) This answer is wrong because it includes statement II, which is false. A ductile material should be designed for use with its maximum stress *lower than* its yield strength.

SOLUTION 9

From a table of properties of standard steel structural sections, the weight of steel for 1.5 in square stock is

$$w = 0.6375 \text{ lbf/in}$$

The force on each cantilever beam is

$$F = wL = \left(0.6375 \ \frac{\text{lbf}}{\text{in}}\right)(48 \text{ in})$$
$$= 30.6 \text{ lbf}$$

The torque on each cantilever beam is

$$T = F\left(\frac{L}{2}\right) = (30.6 \text{ lbf})\left(\frac{48 \text{ in}}{2}\right)$$
$$= 734.4 \text{ in-lbf}$$

A square member's torsion constant (equivalent to an effective polar moment of inertia) is a function of its side.

$$\kappa = 0.1406 a^4$$
$$= (0.1406)(1.5 \text{ in})^4$$
$$= 0.7118 \text{ in}^4/\text{rad}$$

The shear modulus of steel is

$$G = 11.5 \times 10^6 \text{ psi}$$

The angle of twist produced by the first welded piece is

$$\theta_1 = \frac{Td_1}{\kappa G}$$
$$= \frac{(734.4 \text{ in-lbf})(6 \text{ in})}{\left(0.7118 \ \frac{\text{in}^4}{\text{rad}}\right)\left(11.5 \times 10^6 \ \frac{\text{lbf}}{\text{in}^2}\right)}$$
$$= 0.0005383 \text{ rad}$$

The angle of twist produced by the second welded piece is

$$\theta_2 = \frac{Td_2}{\kappa G}$$
$$= \frac{(734.4 \text{ in-lbf})(18 \text{ in})}{\left(0.7118 \ \frac{\text{in}^4}{\text{rad}}\right)\left(11.5 \times 10^6 \ \frac{\text{lbf}}{\text{in}^2}\right)}$$
$$= 0.0016155 \text{ rad}$$

The total angle of twist is

$$\theta_{\text{total}} = \theta_1 + \theta_2$$
$$= 0.0005383 \text{ rad} + 0.001077 \text{ rad}$$
$$= 0.0021538 \text{ rad}$$

The elevation of the second weldment bar is lowered due to angular twist by

$$h = L \sin \theta_{\text{total}}$$
$$= (48 \text{ in}) \sin 0.0021538 \text{ rad}$$
$$= 0.10338 \text{ in} \quad (0.10 \text{ in})$$

The answer is (B).

Why Other Options Are Wrong

(A) This incorrect solution results when the twist from the second weldment section does not take into account the distance between the wall and the first weldment section.

(C) This incorrect solution results when the polar moment of inertia, instead of the torsion constant, is used.

(D) This incorrect solution results when the entire length of the welded arms is used to compute the torque.

SOLUTION 10

The shear modulus of stainless steel is

$$G = 10.6 \times 10^6 \text{ psi}$$

The torsion constant (or effective polar moment of inertia) for an ellipse is

$$\kappa = \frac{\pi a^3 b^3}{a^2 + b^2} = \frac{\pi \left(\frac{0.50 \text{ in}}{2}\right)^3 \left(\frac{0.25 \text{ in}}{2}\right)^3}{\left(\frac{0.50 \text{ in}}{2}\right)^2 + \left(\frac{0.25 \text{ in}}{2}\right)^2}$$
$$= 0.001227 \text{ in}^4/\text{rad}$$

The angle of twist is

$$\theta = \frac{TL}{\kappa G}$$
$$= \left(\frac{(250 \text{ in-lbf})(6 \text{ in})}{\left(0.001227 \frac{\text{in}^4}{\text{rad}}\right)\left(10.6 \times 10^6 \frac{\text{lbf}}{\text{in}^2}\right)}\right)\left(\frac{360°}{2\pi \text{ rad}}\right)$$
$$= 6.6°$$

The shear stress is

$$\tau = \frac{2T}{\pi a b^2} = \frac{(2)(250 \text{ in-lbf})}{\pi \left(\frac{0.50 \text{ in}}{2}\right)\left(\frac{0.25 \text{ in}}{2}\right)^2}$$
$$= 40,744 \text{ psi}$$

The maximum allowable shear for stainless steel is calculated from the yield strength.

$$S_{ys} = \frac{S_{yt}}{\sqrt{3}} = \frac{30,000 \frac{\text{lbf}}{\text{in}^2}}{\sqrt{3}}$$
$$= 17,320 \text{ psi}$$

The shear stress applied to the key is greater than the maximum shear strength for stainless steel, so the driver will fail.

The answer is (D).

Why Other Options Are Wrong

(A) This incorrect solution results when torsion and shear are confused.

(B) This solution results when the full length of the major and minor axes, instead of the semimajor and semiminor distances, are used.

(C) This incorrect solution results from reversing the height and width of the ellipse when calculating the shear.

SOLUTION 11

The moment about a hinge is zero. The part of the bench to the right of the hinge can be isolated. The free-body diagram of this section is

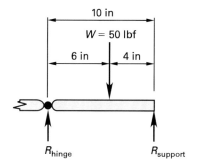

The reaction force at the support can be found from the sum of moments about the hinge.

$$\sum M_{\text{hinge}} = 0 \text{ in-lbf}$$
$$(50 \text{ lbf})(6 \text{ in}) - R_{\text{support}}(10 \text{ in}) = 0 \text{ in-lbf}$$
$$R_{\text{support}} = 30 \text{ lbf}$$

The reaction force at the hinge can be found from the sum of the forces.

$$\sum F = 0 \text{ lbf}$$
$$50 \text{ lbf} - 30 \text{ lbf} - R_{\text{hinge}} = 0 \text{ lbf}$$
$$R_{\text{hinge}} = 20 \text{ lbf}$$

The answer is (B).

Why Other Options Are Wrong

(A) This incorrect solution results when the force at the hinge is determined by inspection to be zero. The lack of support or moment often leads to this misconception.

(C) This incorrect solution results when the reaction forces are assumed to be equal.

(D) This incorrect solution results when the moment is taken from the left along the entire hinge. Two equations will have six unknowns and remain unsolvable.

SOLUTION 12

The normal force of the beam is

$$F_N = mg$$
$$= (15 \text{ slugs})\left(\frac{1 \frac{\text{lbf-sec}^2}{\text{ft}}}{1 \text{ slug}}\right)\left(32.2 \frac{\text{ft}}{\text{sec}^2}\right)$$
$$= 483 \text{ lbf}$$

The friction force of the beam is

$$F_{\text{load}} = fF_N$$
$$= (0.1)(483 \text{ lbf})$$
$$= 48.3 \text{ lbf}$$

The diameter of the wheel is found using the sum of moments.

$$\sum M = 0 \text{ in-lbf}$$
$$F_{\text{load}} d_{\text{pinion}} - F_{\text{applied}} d_{\text{wheel}} = 0 \text{ in-lbf}$$
$$d_{\text{wheel}} = \frac{F_{\text{load}} d_{\text{pinion}}}{F_{\text{applied}}}$$
$$= \frac{(48.3 \text{ lbf})(1.25 \text{ in})}{5 \text{ lbf}}$$
$$= 12.1 \text{ in} \quad (12 \text{ in})$$

The answer is (C).

Why Other Options Are Wrong

(A) This incorrect solution results when the mass of the beam is used as the normal force.

(B) This incorrect solution results when the radius is computed instead of the diameter.

(D) This incorrect solution results when the coefficient of friction is neglected.

SOLUTION 13

The first moments of the areas for the first two spans are

$$A_1 a = \frac{wL^4}{24}$$
$$= \frac{\left(10 \ \frac{\text{lbf}}{\text{ft}}\right)(10 \text{ ft})^4}{24}$$
$$= 4166.67 \text{ lbf-ft}^3$$

$$A_2 b = \frac{FL^3}{16}$$
$$= \frac{(15 \text{ lbf})(10 \text{ ft})^3}{16}$$
$$= 937.5 \text{ lbf-ft}^3$$

Using the three-moment equation with M_1 set to zero, the relationship between M_2 and M_3 for the first span is

$$M_1 L_1 + 2M_2(L_1 + L_2) + M_3 L_2$$
$$= -6\left(\frac{A_1 a}{L_1} + \frac{A_2 b}{L_2}\right)$$
$$0 + 2M_2(10 \text{ ft} + 10 \text{ ft}) + M_3(10 \text{ ft})$$
$$= (-6)\left(\frac{4166.67 \text{ ft}^3\text{-lbf}}{10 \text{ ft}} + \frac{937.5 \text{ ft}^3\text{-lbf}}{10 \text{ ft}}\right)$$
$$40M_2 + 10M_3 = -3062.5026 \text{ ft-lbf}$$

The first moments of the areas for the second and third span are

$$A_2 a = A_2 b = 937.5 \text{ ft}^3\text{-lbf}$$
$$A_3 b = \tfrac{1}{6} Fd(L^2 - d^2)$$
$$= \left(\tfrac{1}{6}\right)(500 \text{ lbf})(3 \text{ ft})\left((5 \text{ ft})^2 - (3 \text{ ft})^2\right)$$
$$= 4000 \text{ ft}^3\text{-lbf}$$

Using the three-moment equation with M_4 set to zero, the relationship between M_2 and M_3 for the second span is

$$M_2 L_2 + 2M_3(L_2 + L_3) + M_4 L_3$$
$$= -6\left(\frac{A_2 a}{L_2} + \frac{A_3 b}{L_3}\right)$$
$$M_2(10 \text{ ft}) + 2M_3(10 \text{ ft} + 5 \text{ ft}) + 0$$
$$= (-6)\left(\frac{937.5 \text{ ft}^3\text{-lbf}}{10 \text{ ft}} + \frac{4000 \text{ ft}^3\text{-lbf}}{5 \text{ ft}}\right)$$
$$10M_2 + 30M_3 = -5362.5 \text{ ft-lbf}$$
$$M_2 = -3M_3 - 536.25 \text{ ft-lbf}$$

Substituting this value for M_2 into the earlier equation and solving for the moment around point B,

$$(40)(-3M_3 - 536.25 \text{ ft-lbf}) + 10M_3$$
$$= -3062.5 \text{ ft-lbf}$$
$$-120M_3 - 21{,}450 \text{ ft-lbf} + 10M_3$$
$$= -3062.5 \text{ ft-lbf}$$

$$-110M_3 = 18387.5 \text{ ft-lbf}$$
$$M_3 = -167.16 \text{ ft-lbf} \quad (-170 \text{ ft-lbf})$$

The answer is (B).

Why Other Options Are Wrong

(A) This incorrect solution results when conventional moment equations are used and the reaction forces are neglected.

(C) This incorrect solution results when the moment around point A is calculated.

(D) This incorrect solution results when the conventional moment equations are used and the reaction forces are considered. The equations will have too many unknowns to be solved.

SOLUTION 14

Two ropes support weight A. Four ropes support weight B. Weight C is supported by one rope. The relationship between the acceleration of the weights is thus

$$2a_A + 4a_B + a_C = 0 \text{ ft/sec}^2$$

The acceleration of weights A and C is $-g/2$.

$$2\left(\frac{-g}{2}\right) + 4a_B + \frac{-g}{2} = 0 \text{ ft/sec}^2$$

Solving for the acceleration of weight B,

$$a_B = \frac{1.5g}{4}$$

$$= \frac{(1.5)\left(32.2 \dfrac{\text{ft}}{\text{sec}^2}\right)}{4}$$

$$= 12.075 \text{ ft/sec}^2 \quad (12 \text{ ft/sec}^2)$$

The answer is (B).

Why Other Options Are Wrong

(A) This incorrect solution results when the acceleration of blocks A and C is taken as $g/2$ instead of $-g/2$.

(C) This incorrect solution results when only two ropes are counted as supports for block B.

(D) This incorrect solution results when the sum of the accelerations is made equal a_B instead of zero.

SOLUTION 15

The tangential velocity is

$$v_t = nC$$
$$= \left(30 \dfrac{\text{rev}}{\text{min}}\right)\left(\dfrac{\pi(18 \text{ in})}{1 \text{ rev}}\right)\left(\dfrac{1 \text{ ft}}{12 \text{ in}}\right)\left(\dfrac{1 \text{ min}}{60 \text{ sec}}\right)$$
$$= 2.36 \text{ ft/sec}$$

The slack-side tension is found from the equation

$$\frac{F_{\max} - \dfrac{mv^2}{g_c}}{F_{\min} - \dfrac{mv^2}{g_c}} = e^{f\theta}$$

$$\frac{mv^2}{g_c} = \frac{\left(5 \dfrac{\text{lbm}}{\text{ft}}\right)\left(2.36 \dfrac{\text{ft}}{\text{sec}}\right)^2}{32.2 \dfrac{\text{lbm-ft}}{\text{lbf-sec}^2}}$$

$$= 0.865 \text{ lbf}$$

$$\frac{225 \text{ lbf} - 0.865 \text{ lbf}}{F_{\min} - 0.865 \text{ lbf}} = e^{(0.37)(60°)\left(\frac{2\pi}{360°}\right)}$$

$$F_{\min} = 153 \text{ lbf} \quad (150 \text{ lbf})$$

The answer is (B).

Why Other Options Are Wrong

(A) This incorrect solution results when the angle of contact with the pulley is not converted into radians from degrees.

(C) This incorrect solution occurs when the gravitational constant, g_c, is neglected while calculating the minimum force.

(D) This incorrect solution results when the speed in rpm instead of the tangential velocity in ft/sec is used.

SOLUTION 16

The area moment of inertia for the flywheel is

$$J = \frac{\pi r^4}{2} = \frac{\pi\left(\dfrac{d}{2}\right)^4}{2}$$

$$= \frac{\pi\left(\dfrac{3 \text{ in}}{2}\right)^4}{2}$$

$$= 7.952 \text{ in}^4$$

The area of the flywheel is

$$A = \pi r^2 = \pi\left(\dfrac{d}{2}\right)^2$$

$$= \pi\left(\dfrac{3 \text{ in}}{2}\right)^2$$

$$= 7.07 \text{ in}^2$$

The polar radius of gyration is

$$r = \sqrt{\frac{J}{A}} = \sqrt{\frac{7.952 \text{ in}^4}{7.07 \text{ in}^2}}$$
$$= 1.06 \text{ in}$$

The work is

$$W = \frac{mr^2\omega^2}{2g_c}$$

$$= \frac{(0.5 \text{ lbm})(1.06 \text{ in})^2 \times \left(\left(72 \frac{\text{rev}}{\text{min}}\right)\left(\frac{1 \text{ min}}{60 \text{ sec}}\right)\left(\frac{2\pi \text{ rad}}{\text{rev}}\right)\right)^2}{(2)\left(32.2 \frac{\text{ft-lbm}}{\text{lbf-sec}^2}\right)\left(12 \frac{\text{in}}{\text{ft}}\right)}$$

$$= 0.04133 \text{ in-lbf} \quad (0.041 \text{ in-lbf})$$

The answer is (A).

Why Other Options Are Wrong

(B) This incorrect solution results when power, instead of work, is computed, and the horsepower conversion factor (63,025) is neglected.

(C) This incorrect solution results when the gravitational constant is neglected.

(D) This incorrect solution results when no conversion factors are used.

SOLUTION 17

The spring constant can be found using the conservation of energy.

$$E_{\text{spring}} = E_{\text{potential}} + E_{\text{kinetic}}$$
$$\frac{1}{2}kx^2 = \frac{mgh}{g_c} + \frac{mv^2}{2g_c}$$

Weight, W, is a function of mass and gravities.

$$W = \frac{mg}{g_c}$$

Substituted, the energy equation becomes

$$\frac{1}{2}kx^2 = Wh + \frac{Wv^2}{2g}$$

The change in height, h, is

$$h = h_2 - h_1$$
$$= (3 \text{ in} - (\cos 45°)(3 \text{ in}))\left(\frac{1 \text{ ft}}{12 \text{ in}}\right)$$
$$= 0.073 \text{ ft}$$

The velocity of the large ball, v, is found using the conservation of momentum.

$$m_{\text{large}} v_{\text{large}} + m_{\text{small}} v_{\text{small}} = m_{\text{large}} v'_{\text{large}} + m_{\text{small}} v'_{\text{small}}$$

$$(5 \text{ oz})\left(\frac{1 \text{ lbf}}{16 \text{ oz}}\right)v_{\text{large}} + (3 \text{ oz})\left(\frac{1 \text{ lbf}}{16 \text{ oz}}\right)\left(\frac{0 \text{ ft}}{\text{sec}}\right)$$
$$= (5 \text{ oz})\left(\frac{1 \text{ lbf}}{16 \text{ oz}}\right)\left(\frac{1.5 \text{ ft}}{\text{sec}}\right) + (3 \text{ oz})\left(\frac{1 \text{ lbf}}{16 \text{ oz}}\right)\left(\frac{4 \text{ ft}}{\text{sec}}\right)$$
$$v_{\text{large}} = 3.9 \text{ ft/sec}$$

Solving the energy equation, k is

$$(5 \text{ oz})\left(\frac{1 \text{ lbf}}{16 \text{ oz}}\right)(0.073 \text{ ft})$$
$$+ \frac{(5 \text{ oz})\left(\frac{1 \text{ lbf}}{16 \text{ oz}}\right)\left(\frac{3.9 \text{ ft}}{\text{sec}}\right)^2}{(2)\left(\frac{32.2 \text{ ft}}{\text{sec}^2}\right)}$$
$$= \frac{1}{2}k\left((1 \text{ in})\left(\frac{1 \text{ ft}}{12 \text{ in}}\right)\right)^2$$
$$k = 27.84 \text{ lbf/ft} \quad (28 \text{ lbf/ft})$$

The answer is (B).

Why Other Options Are Wrong

(A) This incorrect solution occurs when the velocity of the small ball is used in the conservation of energy calculation.

(C) This incorrect solution occurs when the length of the pendulum, 3 in, is used for the height, h.

(D) This incorrect solution occurs when the conversion from ounces to pound force is ignored.

SOLUTION 18

The velocity of the pendulum at its maximum recoil height is

$$\begin{aligned}
v_{pend} &= \sqrt{2gh} \\
&= \sqrt{(2)\left(32.2 \ \frac{ft}{sec^2}\right)(18 \ in)\left(\frac{1 \ ft}{12 \ in}\right)} \\
&= 9.8285 \ ft/sec
\end{aligned}$$

The required mass of the pendulum is found from the conservation of momentum.

$$m_{bullet}v_{bullet} = (m_{bullet} + m_{pend})v_{pend}$$

$$\begin{aligned}
m_{pend} &= \frac{m_{bullet}v_{bullet}}{v_{pend}} - m_{bullet} \\
&= \frac{(80 \ gr)\left(1800 \ \frac{ft}{sec}\right)}{9.8285 \ \frac{ft}{sec}} - 80 \ gr \\
&\quad \frac{}{7000 \ \frac{gr}{lbm}} \\
&= 2.0816 \ lbm \quad (2.1 \ lbm)
\end{aligned}$$

The answer is (C).

Why Other Options Are Wrong

(A) This incorrect solution results when the square root is neglected during the calculation of the pendulum's velocity.

(B) This incorrect solution results when conversion factors are neglected.

(D) This solution results when grains are incorrectly equated with grams.

SOLUTION 19

stage	angular acceleration (rad/sec²)	speed (rpm)
1	0.25	0–10
2	0.125	10–15
3	0.375	15–45

The angular velocity at the beginning and end of each stage is

$$\omega_1 = 0$$

$$\begin{aligned}
\omega_2 &= \left(10 \ \frac{rev}{min}\right)\left(2\pi \ \frac{rad}{rev}\right)\left(\frac{1 \ min}{60 \ sec}\right) \\
&= 1.0472 \ rad/sec
\end{aligned}$$

$$\begin{aligned}
\omega_3 &= \left(15 \ \frac{rev}{min}\right)\left(2\pi \ \frac{rad}{rev}\right)\left(\frac{1 \ min}{60 \ sec}\right) \\
&= 1.5708 \ rad/sec
\end{aligned}$$

$$\begin{aligned}
\omega_4 &= \left(45 \ \frac{rev}{min}\right)\left(2\pi \ \frac{rad}{rev}\right)\left(\frac{1 \ min}{60 \ sec}\right) \\
&= 4.7124 \ rad/sec
\end{aligned}$$

The time it takes to complete each step is

$$\begin{aligned}
\Delta t_1 &= \frac{\omega_2 - \omega_1}{\alpha_1} \\
&= \frac{1.0472 \ \frac{rad}{sec} - 0 \ \frac{rad}{sec}}{0.25 \ \frac{rad}{sec^2}} \\
&= 4.1888 \ sec
\end{aligned}$$

$$\begin{aligned}
\Delta t_2 &= \frac{\omega_3 - \omega_2}{\alpha_2} \\
&= \frac{1.5708 \ \frac{rad}{sec} - 1.0472 \ \frac{rad}{sec}}{0.125 \ \frac{rad}{sec^2}} \\
&= 4.1888 \ sec
\end{aligned}$$

$$\begin{aligned}
\Delta t_3 &= \frac{\omega_4 - \omega_3}{\alpha_3} \\
&= \frac{4.7124 \ \frac{rad}{sec} - 1.5708 \ \frac{rad}{sec}}{0.375 \ \frac{rad}{sec^2}} \\
&= 8.3776 \ sec
\end{aligned}$$

Acceleration is steady within each stage, so the number of revolutions during each step is

$$\begin{aligned}
N_1 &= \frac{1}{2}\Delta t_1(\omega_2 + \omega_1) \\
&= \left(\frac{1}{2}\right)(4.1888 \ sec)\left(1.0472 \ \frac{rad}{sec} + 0 \ rad\right) \\
&\quad \times \left(\frac{1 \ rev}{2\pi \ rad}\right) \\
&= 0.3491 \ rev
\end{aligned}$$

$$N_2 = \frac{1}{2}\Delta t_2(\omega_3 + \omega_2)$$

$$= \left(\frac{1}{2}\right)(4.1888 \text{ sec})\left(1.5708 \frac{\text{rad}}{\text{sec}} + 1.0472 \frac{\text{rad}}{\text{sec}}\right)$$

$$\times \left(\frac{1 \text{ rev}}{2\pi \text{ rad}}\right)$$

$$= 0.8726 \text{ rev}$$

$$N_3 = \frac{1}{2}\Delta t_3(\omega_4 + \omega_3)$$

$$= \left(\frac{1}{2}\right)(8.3776 \text{ sec})\left(4.7124 \frac{\text{rad}}{\text{sec}} + 1.5708 \frac{\text{rad}}{\text{sec}}\right)$$

$$\times \left(\frac{1 \text{ rev}}{2\pi \text{ rad}}\right)$$

$$= 4.1888 \text{ rev}$$

The total number of revolutions is

$$N_{\text{total}} = N_1 + N_2 + N_3$$
$$= 0.3491 \text{ rev} + 0.8726 \text{ rev} + 4.1888 \text{ rev}$$
$$= 5.4105 \text{ rev}$$

The total amount of material used during the ramp-up is

$$m = N_{\text{total}} \dot{m} \omega_0$$

$$= (5.4105 \text{ rev})\left(400 \frac{\text{lbm}}{\text{hr}}\right)\left(\frac{1 \text{ min}}{40 \text{ rev}}\right)$$

$$\times \left(\frac{1 \text{ hr}}{60 \text{ min}}\right)\left(0.4536 \frac{\text{kg}}{\text{lbm}}\right)$$

$$= 0.41 \text{ kg}$$

The answer is (B).

Why Other Options Are Wrong

(A) This incorrect solution results when only the final speed of the extruder is used for calculations.

(C) This incorrect solution results when the total time and mass flow rate are used to compute the material quantity. This method can be used for a steady-state system with a constant flow rate, but in this system the extruder is accelerating.

(D) This incorrect solution results when the reference extruder speed is left out of the mass equation.

SOLUTION 20

The force at the maximum deflection is

$$F = 1000 \text{ lbf} - \left(50 \frac{\text{lbf}}{\text{deg}}\right)(10°)$$
$$= 500 \text{ lbf}$$

The modulus of elasticity for steel is

$$E = 30 \times 10^6 \text{ psi}$$

The moment of inertia is found from the formula

$$y_{\text{max}} = -\frac{FL^3}{3EI}$$

$$I = -\frac{FL^3}{3Ey_{\text{max}}}$$

$$= -\frac{(500 \text{ lbf})(1.5 \text{ in})^3}{(3)\left(30 \times 10^6 \frac{\text{lbf}}{\text{in}^2}\right)(-0.01 \text{ in})}$$

$$= 0.001875 \text{ in}^4$$

The diameter is found from the formula

$$I = \frac{\pi \left(\frac{d}{2}\right)^4}{4}$$

$$d = 2\left(\frac{4I}{\pi}\right)^{1/4}$$

$$= (2)\left(\frac{(4)(0.001875 \text{ in}^4)}{\pi}\right)^{1/4}$$

$$= 0.4420 \text{ in} \quad (0.50 \text{ in})$$

The answer is (C).

Why Other Options Are Wrong

(A) This incorrect solution results when the dowel is treated as a column.

(B) This incorrect solution results when the rating of the load cell, instead of the force applied by the load cell on the dowel, is used.

(D) This incorrect solution results when the total length of the dowel is used. The full length of the dowel is inconsequential. Only the length between the fixed end of the dowel and the point of contact with the cell matters.

SOLUTION 21

The moment at the center of the beam (including the added weight) is

$$M = \frac{PL}{4}$$
$$= \frac{(1000 \text{ lbf} + 2500 \text{ lbf})(12 \text{ ft})}{4}$$
$$= 10{,}500 \text{ ft-lbf}$$

The required section modulus of the additional steel is

$$S = \frac{M}{\sigma}$$
$$= \left(\frac{10{,}500 \text{ ft-lbf}}{24{,}000 \frac{\text{lbf}}{\text{in}^2}}\right)\left(12 \frac{\text{in}}{\text{ft}}\right)$$
$$= 5.25 \text{ in}^3$$

The section modulus of a rectangular section measuring $b \times h$ is

$$S = \frac{bh^2}{6}$$
$$5.25 \text{ in}^3 = \frac{b(9 \text{ in})^2}{6}$$
$$b = 0.389 \text{ in}$$

The number of 0.25 in plies needed to meet the width requirement is

$$N_{\text{plates}} = \frac{b}{t}$$
$$= \frac{0.389 \text{ in}}{0.25 \text{ in}}$$
$$= 1.56 \quad (2)$$

The answer is (B).

Why Other Options Are Wrong

(A) This incorrect solution results when the conversion from feet to inches is neglected in the section modulus.

(C) This incorrect solution results when the moment equation is divided by 2 instead of 4.

(D) This incorrect solution results when the moment equation is not divided by 4.

SOLUTION 22

The section modulus of the beam before damage was

$$S_0 = \frac{bh^2}{6}$$
$$= \frac{(6 \text{ in})(8 \text{ in})^2}{6}$$
$$= 64 \text{ in}^3$$

The section modulus at the damaged section of the beam is

$$S_d = \frac{bh^2}{6}$$
$$= \frac{(6 \text{ in})(8 \text{ in} - 2 \text{ in})^2}{6}$$
$$= 36 \text{ in}^3$$

The maximum moment on a simply supported beam carrying a distributed load occurs at the midpoint.

$$M_0 = \frac{wL^2}{8} = \sigma_a S_0$$

The moment at the damaged section of the beam is

$$M_d = \left(\frac{w_0}{2}\right)(Lx - x^2)$$
$$= \sigma_a S_d$$

The maximum allowable distributed load before the damage can be found from the equation

$$w_0 = \frac{8\sigma_a S_0}{L^2}$$
$$= \frac{(8)\left(1200 \frac{\text{lbf}}{\text{in}^2}\right)(64 \text{ in}^3)}{(10 \text{ ft})^2 \left(12 \frac{\text{in}}{\text{ft}}\right)^2}$$
$$= 42.67 \text{ lbf/in}$$

If the damaged beam was loaded to the undamaged beam's capacity, the bending stress at the damaged location would be

$$\sigma = \frac{M}{S_d} = \frac{\left(\frac{w_0}{2}\right)(Lx - x^2)}{S_d}$$

$$= \frac{\left(\frac{42.67 \frac{\text{lbf}}{\text{in}}}{2}\right)((10 \text{ ft})(4 \text{ ft}) - (4 \text{ ft})^2)\left(12 \frac{\text{in}}{\text{ft}}\right)^2}{36 \text{ in}^3}$$

$$= 2048 \text{ lbf/in}^2$$

Since bending stress is proportional to the distributed load, w, the reduction in distributed load needed to reduce the bending stress to 1200 psi is

$$\frac{2048 \frac{\text{lbf}}{\text{in}^2} - 1200 \frac{\text{lbf}}{\text{in}^2}}{2048 \frac{\text{lbf}}{\text{in}^2}} = 0.414 \quad (41.4\%)$$

The answer is (B).

Why Other Options Are Wrong

(A) This incorrect solution results when the reduction is calculated from the change in physical dimensions of the beam.

(C) This incorrect answer results from not recognizing that the stress at the beam's midpoint will exceed 1200 psi even though the stress at the damaged section is limited to 1200 psi.

(D) This incorrect solution results when the percentage is calculated to determine the remaining useful percentage of the beam rather than the reduction.

SOLUTION 23

The actual dimensions of a 2 × 12 joist are approximately 1.5 in and 11.25 in. The actual thickness of a 1 in floor board is approximately 0.75 in.

Calculate the self-weight loading of each cantilevered joist and its associated contributory flooring.

$$F_{\text{deck}} = (b_{\text{joist}} h_{\text{joist}} L_{\text{joist}} + b_{\text{floor}} h_{\text{floor}} L_{\text{floor}})\gamma_{\text{wood}}$$

$$= \begin{pmatrix} (1.5 \text{ in})(11.25 \text{ in})(80 \text{ in}) \\ + (0.75 \text{ in})(16 \text{ in})(80 \text{ in}) \end{pmatrix}$$

$$\times \left(35 \frac{\text{lbf}}{\text{ft}^3}\right)\left(\frac{1 \text{ ft}}{12 \text{ in}}\right)^3$$

$$= 46.79 \text{ lbf}$$

The end moment on the cantilevered joist is

$$M_{\text{dead}} = \frac{F_{\text{deck}} L_{\text{joist}}}{2}$$

$$= \frac{(46.79 \text{ lbf})(80 \text{ in})}{2}$$

$$= 1872 \text{ in-lbf}$$

The moment created by the live load of the plastic compound hopper is

$$M_{\text{live}} = \frac{F_{\text{hopper}} L_{\text{joist}}}{2}$$

$$= \frac{(500 \text{ lbf})(80 \text{ in})}{2}$$

$$= 20{,}000 \text{ in-lbf}$$

The total moment is

$$M_{\text{total}} = M_{\text{dead}} + M_{\text{live}}$$

$$= 1872 \text{ in-lbf} + 20{,}000 \text{ in-lbf}$$

$$= 21{,}872 \text{ in-lbf}$$

The section modulus of each joist is

$$S_{\text{joist}} = \frac{bh^2}{6}$$

$$= \frac{(1.5 \text{ in})(11.25 \text{ in})^2}{6}$$

$$= 31.6 \text{ in}^3$$

The bending stress on each joist is

$$\sigma = \frac{M_{\text{total}}}{S_{\text{joist}}}$$

$$= \frac{21{,}872 \text{ in-lbf}}{31.6 \text{ in}^3}$$

$$= 692 \text{ lbf/in}^2 \quad (690 \text{ psi})$$

The answer is (C).

Why Other Options Are Wrong

(A) This incorrect solution results when the moment of inertia is used instead of the section modulus.

(B) This incorrect solution results when the weight of the platform and safety railing is neglected.

(D) This incorrect solution results when the full length is used to calculate the moment.

SOLUTION 24

A force fit (or shrink fit) is a special type of interference fit. A force fit is characterized by constant bore pressure throughout the range of sizes.

The answer is (C).

Why Other Options Are Wrong

(A) A running and sliding fit would not apply significant pressure to the bearing, allowing it to slip in place. A running and sliding fit is intended for accurate location with lubrication allowance.

(B) A locational fit would determine an accurate location of the bearing, but not a constant pressure. A locational fit is intended to determine only the location of the mating parts.

(D) "Accurate fit" is not a class of fit. The classes of fits are arranged in three general groups: running and sliding fits, locational fits, and force fits.

SOLUTION 25

The total number of revolutions during the rated life of the bearing is

$$N_{\text{rev}} = L_R n_b$$
$$= (3000 \text{ hr})\left(500 \ \frac{\text{rev}}{\text{min}}\right)\left(60 \ \frac{\text{min}}{\text{hr}}\right)$$
$$= 90 \times 10^6 \text{ rev}$$

The speed of the winding mechanism is

$$n_m = \frac{\text{v}}{C}$$
$$= \frac{300 \ \frac{\text{ft}}{\text{min}}}{0.20 \text{ ft}}$$
$$= 1500 \text{ rev/min}$$

The design life of the bearing in machine hours is

$$L_D = \frac{N_{\text{rev}}}{n_m}$$
$$= \left(\frac{90 \times 10^6 \text{ rev}}{1500 \ \frac{\text{rev}}{\text{min}}}\right)\left(\frac{1 \text{ hr}}{60 \text{ min}}\right)$$
$$= 1000 \text{ hr}$$

The answer is (C).

Why Other Options Are Wrong

(A) This incorrect solution results from not converting the catalog-rated life of the bearing from hours to minutes.

(B) This incorrect solution results from confusing the catalog-rated speed with the speed of the winding mechanism.

(D) This incorrect solution results from using the catalog-rated life of the bearing instead of the required design life.

SOLUTION 26

Calculate the dynamic equivalent radial load for each phase. (The rotation factor, V, equals 1 for a rotating inner race and 1.2 for a rotating outer race.)

$$F = VXF_R + YF_T$$
$$F_1 = (1)(0.4)(250 \text{ lbf}) + (0.4 \cot 20°)(500 \text{ lbf})$$
$$= 649 \text{ lbf}$$
$$F_2 = (1)(0.4)(750 \text{ lbf}) + (0.4 \cot 20°)(1000 \text{ lbf})$$
$$= 1399 \text{ lbf}$$
$$F_3 = (1)(0.4)(250 \text{ lbf}) + (0.4 \cot 20°)(500 \text{ lbf})$$
$$= 649 \text{ lbf}$$

The mean load is solved by using Miner's rule, combining and averaging individual loads with exponent p taken as 10/3 for roller bearings.

$$F_{\text{mean}} = \left(\frac{\sum F_i^p n_i t_i}{\sum n_i t_i}\right)^{1/p}$$
$$= \left(\frac{F_1^{10/3} n_1 t_1 + F_2^{10/3} n_2 t_2 + F_3^{10/3} n_3 t_3}{n_1 t_1 + n_2 t_2 + n_3 t_3}\right)^{3/10}$$

Convert rotational speed to revolutions per second.

$$n_1 = \left(500 \ \frac{\text{rev}}{\text{min}}\right)\left(\frac{1 \text{ min}}{60 \text{ sec}}\right)$$
$$= 8.3 \text{ rev/sec}$$
$$n_2 = \left(5000 \ \frac{\text{rev}}{\text{min}}\right)\left(\frac{1 \text{ min}}{60 \text{ sec}}\right)$$
$$= 83.3 \text{ rev/sec}$$
$$n_3 = \left(250 \ \frac{\text{rev}}{\text{min}}\right)\left(\frac{1 \text{ min}}{60 \text{ sec}}\right)$$
$$= 4.2 \text{ rev/sec}$$

$$F_{\text{mean}} = \left[\frac{\begin{array}{c}(649\ \text{lbf})^{10/3}\left(8.3\ \dfrac{\text{rev}}{\text{sec}}\right)(7\ \text{sec}) \\ +(1399\ \text{lbf})^{10/3}\left(83.3\ \dfrac{\text{rev}}{\text{sec}}\right) \\ \times (10\ \text{sec}) \\ +(649\ \text{lbf})^{10/3}\left(4.2\ \dfrac{\text{rev}}{\text{sec}}\right)(3\ \text{sec}) \end{array}}{\begin{array}{c}\left(8.3\ \dfrac{\text{rev}}{\text{sec}}\right)(7\ \text{sec}) + \left(83.3\ \dfrac{\text{rev}}{\text{sec}}\right) \\ \times (10\ \text{sec}) \\ +\left(4.2\ \dfrac{\text{rev}}{\text{sec}}\right)(3\ \text{sec})\end{array}} \right]^{3/10}$$

$$= 1368\ \text{lbf} \quad (1400\ \text{lbf})$$

The answer is (C).

Why Other Options Are Wrong

(A) This incorrect answer results when the mean load is obtained by simply averaging the three loads.

(B) This incorrect answer results when the exponents p and $1/p$ are neglected in Miner's rule.

(D) This incorrect answer results when the conversion from rev/min to rev/sec is neglected.

SOLUTION 27

The relation constant between respective loads and lives for ball bearings is

$$a = 3$$

The reliability for this application is found from the equation

$$F_R = F_D \left(\frac{\dfrac{L_D n_D}{L_R n_R}}{0.02 + (4.439)\left(\ln \dfrac{1}{R}\right)^{1/1.483}} \right)^{1/a}$$

$$800\ \text{lbf} = 750\ \text{lbf} \left(\frac{\dfrac{(1000\ \text{hr})\left(1500\ \dfrac{\text{rev}}{\text{min}}\right)}{(3000\ \text{hr})\left(500\ \dfrac{\text{rev}}{\text{min}}\right)}}{0.02 + (4.439)\left(\ln \dfrac{1}{R}\right)^{1/1.483}} \right)^{1/3}$$

$$R = 0.923716 \quad (0.92)$$

The answer is (D).

Why Other Options Are Wrong

(A) This incorrect solution results when the exponent $1/1.483$ is neglected.

(B) This incorrect solution results when the catalog-rated load and the design radial load are reversed.

(C) This incorrect solution results when the relation constant between respective loads and the rated and design lives for ball bearings is neglected.

SOLUTION 28

The friction variable is

$$\left(\frac{r}{c}\right) f = 3.50$$

The coefficient of friction is

$$f = 3.50 \left(\frac{c}{r}\right)$$
$$= (3.50) \left(\frac{0.001\ \text{in}}{\dfrac{1.25\ \text{in}}{2}} \right)$$
$$= 0.0056$$

The frictional torque is

$$T = fWr = f(2rLp)r$$
$$= 2r^2fLp$$
$$= (2)\left(\frac{1.25 \text{ in}}{2}\right)^2(0.0056)(3.0 \text{ in})\left(250\ \frac{\text{lbf}}{\text{in}^2}\right)$$
$$= 3.28125 \text{ in-lbf}$$

The shear stress is

$$\tau = \frac{T}{(2\pi rL)r}$$
$$= \frac{3.28125 \text{ in-lbf}}{2\pi\left(\dfrac{1.25 \text{ in}}{2}\right)^2(3.0 \text{ in})}$$
$$= 0.4456 \text{ lbf/in}^2$$

The absolute viscosity is

$$\mu = \frac{c\tau}{2\pi rn}$$
$$= \frac{(0.001 \text{ in})\left(0.4456\ \dfrac{\text{lbf}}{\text{in}^2}\right)}{2\pi\left(\dfrac{1.25 \text{ in}}{2}\right)\left(25\ \dfrac{\text{rev}}{\text{min}}\right)\left(\dfrac{1 \text{ min}}{60 \text{ sec}}\right)}$$
$$= 2.7 \times 10^{-4} \text{ lbf-sec/in}^2 \quad (2.7 \times 10^{-4} \text{ reyn})$$

The answer is (A).

Why Other Options Are Wrong

(B) This incorrect solution results when the diameter is used instead of the radius.

(C) This incorrect solution results when the torque is used as the shearing stress.

(D) This incorrect solution results when the radius is not squared when calculating the frictional torque.

SOLUTION 29

The coefficient of friction is

$$f = 3.50\left(\frac{c}{r}\right)$$
$$= (3.50)\left(\frac{0.0015 \text{ in}}{\dfrac{1.5 \text{ in}}{2}}\right)$$
$$= 0.007$$

The generated heat in Btu/sec is given by the formula

$$H = 2\pi Tn$$
$$= 2\pi fWrn$$
$$= (2\pi)(0.007)(750 \text{ lbf})\left(\frac{1.5 \text{ in}}{2}\right)\left(\frac{1 \text{ Btu}}{9336 \text{ in-lbf}}\right)$$
$$\quad \times \left(30\ \frac{\text{rev}}{\text{min}}\right)\left(\frac{1 \text{ min}}{60 \text{ sec}}\right)$$
$$= 0.001325 \text{ Btu/sec}$$

The flow rate is

$$Q = \frac{H}{\rho c \Delta T}$$
$$= \left(\frac{0.001325\ \dfrac{\text{Btu}}{\text{sec}}}{\left(0.0311\ \dfrac{\text{lbf}}{\text{in}^3}\right)\left(0.42\ \dfrac{\text{Btu}}{\text{lbf-°F}}\right)(30°\text{F})}\right)\left(\frac{1 \text{ ft}}{12 \text{ in}}\right)^3$$
$$= 1.96 \times 10^{-6} \text{ ft}^3/\text{sec} \quad (2.0 \times 1.0^{-6} \text{ ft}^3/\text{sec})$$

The answer is (B).

Why Other Options Are Wrong

(A) This incorrect solution results when the conversion from in-lbf to Btu is omitted.

(C) This incorrect solution results when the speed is not converted from rev/min to rev/sec.

(D) This incorrect solution results when the flow rate is not converted from cubic inches to cubic feet.

SOLUTION 30

Let clockwise be positive. The speed of the ring around the sun gear is

$$n_6 = n_7$$
$$= n_2\left(\frac{N_2}{N_3}\right)\left(\frac{N_4}{N_5}\right)\left(\frac{N_5}{N_6}\right)$$
$$= \left(250\ \frac{\text{rev}}{\text{min}}\right)\left(\frac{-12}{32}\right)\left(\frac{-14}{36}\right)\left(\frac{-36}{192}\right)$$
$$= -6.8359 \text{ rev/min} \quad [\text{counterclockwise}]$$

The train value of the planetary system is

$$\text{TV} = -\left(\frac{N_7}{N_8}\right)\left(\frac{N_8}{N_9}\right)$$
$$= -\left(\frac{160}{48}\right)\left(\frac{48}{64}\right)$$
$$= -2.5$$

The sun gear is fixed. The speed of the arm is found from the equation

$$\text{TV} = \frac{N_7}{N_9} = \frac{n_9 - n_{\text{arm}}}{n_7 - n_{\text{arm}}}$$

$$n_{\text{arm}} = \frac{(\text{TV})n_7 - n_9}{\text{TV} - 1}$$

$$= \frac{(-2.5)\left(-6.8359 \ \frac{\text{rev}}{\text{min}}\right) - 0 \ \frac{\text{rev}}{\text{min}}}{-2.5 - 1}$$

$$= 4.883 \ \text{rev/min} \quad (4.9 \ \text{rpm})$$

The answer is (A).

Why Other Options Are Wrong

(B) This incorrect answer results from reversing the number of teeth for gears 8 and 9.

(C) This incorrect solution results when the speed of the planet gear is assumed to be the same as that of the ring gear in the planetary arrangement.

(D) This incorrect solution results when the train value is taken as a positive number.

SOLUTION 31

The speed of the pinion is calculated from the equation

$$\frac{d_{\text{gear}}}{d_{\text{pinion}}} = \frac{n_{\text{pinion}}}{n_{\text{gear}}}$$

$$n_{\text{pinion}} = \left(\frac{d_{\text{gear}}}{d_{\text{pinion}}}\right) n_{\text{gear}}$$

$$= \left(\frac{12 \ \text{in}}{6 \ \text{in}}\right)\left(1250 \ \frac{\text{rev}}{\text{min}}\right)$$

$$= 2500 \ \text{rpm}$$

The pitch circle velocity, or tangential velocity, is the same for both gears.

$$v_t = \left(\frac{\pi d_{\text{pinion}}}{1 \ \text{rev}}\right) n_{\text{pinion}}$$

$$= \left(\frac{\pi(6 \ \text{in})\left(\frac{1 \ \text{ft}}{12 \ \text{in}}\right)}{1 \ \text{rev}}\right)\left(2500 \ \frac{\text{rev}}{\text{min}}\right)$$

$$= 3930 \ \text{ft/min}$$

The transmitted load is

$$F_t = \frac{33{,}000 P}{v_t}$$

$$= \frac{\left(33{,}000 \ \frac{\text{ft-lbf}}{\text{hp-min}}\right)(25 \ \text{hp})}{3930 \ \frac{\text{ft}}{\text{min}}}$$

$$= 210 \ \text{lbf}$$

The applied torque is

$$T = \left(\frac{d}{2}\right) F_t$$

$$= \left(\frac{6 \ \text{in}}{2}\right)\left(\frac{1 \ \text{ft}}{12 \ \text{in}}\right)(210 \ \text{lbf})$$

$$= 53 \ \text{ft-lbf}$$

The answer is (A).

Why Other Options Are Wrong

(B) This incorrect solution results when the speed of the pinion, instead of the pitch circle velocity, is used to compute the transmitted load.

(C) This incorrect solution results when the diameter, instead of the radius, is used to calculate the applied torque.

(D) This incorrect solution results when the computation stops at the transmitted load instead of continuing to determine the applied torque.

SOLUTION 32

The number of teeth of the pinion is found by using the equation

$$\tan \Gamma = \frac{N_{\text{gear}}}{N_{\text{pinion}}}$$

$$N_{\text{pinion}} = \frac{N_{\text{gear}}}{\tan \Gamma}$$

$$= \frac{96 \ \text{teeth}}{\tan 70°}$$

$$= 35 \ \text{teeth}$$

The answer is (B).

Why Other Options Are Wrong

(A) This incorrect solution results when the pitch angle of the pinion is calculated.

(C) This incorrect solution results when the arc tangent of the gear pitch angle, instead of the tangent, is used.

(D) This incorrect solution results when the pitch angle of the pinion, instead of the gear, is used.

SOLUTION 33

The relationships among rotational speed, pinion diameter, transmitted load, and power are given by the equations

$$v = \pi d_{\text{pinion}} n_{\text{pinion}}$$
$$P = F_t v$$

Combining them,

$$P = F_t \pi d_{\text{pinion}} n_{\text{pinion}}$$

Multiplying n_{pinion} by 3, d_{pinion} by 0.5, and F_t by 4 causes P to be multiplied by $(3)(0.5)(4)$, or 6, which is equivalent to a 600% increase.

The answer is (D).

Why Other Options Are Wrong

(A) This incorrect solution results from dividing P by 6 instead of multiplying.

(B) This incorrect solution results from dividing the load by the velocity to get the power, instead of multiplying the load and velocity.

(C) This solution results from incorrectly squaring the diameter in the equation for velocity.

SOLUTION 34

The unit spring deflection is the thickness of one plate.

$$\Delta \delta = 0.250 \text{ in}$$

The unit force is the weight of one plate.

$$\Delta F = m\left(\frac{g}{g_c}\right)$$
$$= (4.83 \text{ lbm}) \left(\frac{32.2 \dfrac{\text{ft}}{\text{sec}^2}}{32.2 \dfrac{\text{ft-lbm}}{\text{lbf-sec}^2}} \right)$$
$$= 4.83 \text{ lbf}$$

The equivalent spring constant is

$$k_{\text{eq}} = \frac{\Delta F}{\Delta \delta}$$
$$= \frac{4.83 \text{ lbf}}{0.250 \text{ in}}$$
$$= 19.32 \text{ lbf/in}$$

The spring constant of the four identical springs in parallel is found from

$$k_{\text{eq}} = k_1 + k_2 + k_3 + k_4$$
$$= 4k$$
$$k = \frac{k_{\text{eq}}}{4}$$
$$= \frac{19.32 \dfrac{\text{lbf}}{\text{in}}}{4}$$
$$= 4.83 \text{ lbf/in}$$

The number of active coils in each spring is

$$N_{\text{active}} = N_{\text{total}} - 2$$
$$= 84 - 2$$
$$= 82$$

The shear modulus of stainless steel wire (ASTM A313) is found in a table of spring properties.

$$G = 10.0 \times 10^6 \text{ psi}$$

The mean diameter of the spring is found from the load-deflection equation.

$$k = \frac{G d_{\text{wire}}^4}{8 d_{\text{spring}}^3 N_{\text{active}}}$$
$$d_{\text{spring}} = \left(\frac{G d_{\text{wire}}^4}{8 k N_{\text{active}}} \right)^{1/3}$$
$$= \left(\frac{\left(10.0 \times 10^6 \dfrac{\text{lbf}}{\text{in}^2}\right)(0.187 \text{ in})^4}{(8)\left(4.83 \dfrac{\text{lbf}}{\text{in}}\right)(82)} \right)^{1/3}$$
$$= 1.5686 \text{ in}$$

The inside diameter of the spring is

$$d_i = d_{\text{spring}} - d_{\text{wire}}$$
$$= 1.5686 \text{ in} - 0.187 \text{ in}$$
$$= 1.3816 \text{ in}$$

The nearest smaller standard diameter is 1.25 in.

The answer is (C).

Why Other Options Are Wrong

(A) This incorrect solution results when the formula for springs in series, instead of that for springs in parallel, is used.

(B) This incorrect solution results when the formula for equivalent spring constant in parallel is omitted.

(D) This incorrect solution results when the modulus of elasticity, E, is used instead of the shear modulus, G.

SOLUTION 35

The pitch diameter of the worm gear is

$$d_{\text{worm}} = \frac{L}{\pi \tan \theta} = \frac{4 \text{ in}}{\pi \tan 15°}$$
$$= 4.7518 \text{ in}$$

The pitch diameter of the mating gear is

$$d_{\text{mating}} = 2C - d_{\text{worm}}$$
$$= (2)(21.4 \text{ in}) - 4.7518 \text{ in}$$
$$= 38.048 \text{ in}$$

The number of teeth on the mating gear is

$$N_{\text{mating}} = \frac{\pi d_{\text{mating}}}{p_{\text{mating}}}$$
$$= \frac{\pi(38.048 \text{ in})}{1.25 \text{ in}}$$
$$= 95.625 \quad (96)$$

The answer is (D).

Why Other Options Are Wrong

(A) This incorrect solution results when π is placed in the denominator when calculating the number of teeth on the mating gear.

(B) This incorrect solution results when π is left out of the equation when calculating the number of teeth on the mating gear.

(C) This incorrect solution results when diameter of the worm gear is subtracted from C instead of $2C$.

SOLUTION 36

The torsional spring constant is

$$k_r = \frac{T}{\theta} = \frac{10 \text{ in-lbf}}{\left(\frac{3}{4} \text{ turn}\right)\left(2\pi \frac{\text{rad}}{\text{turn}}\right)}$$
$$= 2.122 \text{ in-lbf/rad} \quad (2 \text{ in-lbf/rad})$$

The answer is (B).

Why Other Options Are Wrong

(A) This incorrect solution results when the stress concentration factor, instead of the torsional spring constant, is calculated.

(C) This incorrect solution results when the spring index is assumed to be the same as the torsional spring constant.

(D) This incorrect solution results when the torque is used as the bending moment for computation.

SOLUTION 37

The time it takes for a water balloon to drop 60 ft is found from the equation

$$h = \frac{1}{2}gt^2$$
$$t = \sqrt{\frac{2h}{g}} = \sqrt{\frac{(2)(60 \text{ ft})}{32.2 \frac{\text{ft}}{\text{sec}^2}}}$$
$$= 1.93 \text{ sec}$$

To travel horizontally 50 ft within 1.93 sec, the velocity of the water balloon as it leaves the launch point must be

$$\text{v} = \frac{R}{t} = \frac{50 \text{ ft}}{1.93 \text{ sec}}$$
$$= 25.9 \text{ ft/sec}$$

The mass of a water balloon is

$$m = \frac{Wg_c}{g} = \frac{(4 \text{ lbf})\left(32.2 \frac{\text{ft-lbm}}{\text{lbf-sec}^2}\right)}{32.2 \frac{\text{ft}}{\text{sec}^2}}$$
$$= 4 \text{ lbm}$$

The kinetic energy that must be given to the water balloon to achieve a velocity of 25.9 ft/sec is

$$E_{\text{kinetic,balloon}} = \frac{m\text{v}^2}{2g_c} = \frac{(4\text{ lbm})\left(25.9\ \dfrac{\text{ft}}{\text{sec}}\right)^2}{(2)\left(32.2\ \dfrac{\text{ft-lbm}}{\text{lbf-sec}^2}\right)}$$
$$= 41.7\text{ ft-lbf}$$

The potential energy stored in each spring is

$$E_{\text{potential,spring}} = \frac{1}{2}k\delta^2 = \frac{\left(\dfrac{1}{2}\right)\left(3\ \dfrac{\text{lbf}}{\text{in}}\right)(6\text{ in})^2}{12\ \dfrac{\text{in}}{\text{ft}}}$$
$$= 4.5\text{ ft-lbf}$$

The minimum number of springs needed is

$$N = \frac{E_{\text{kinetic,balloon}}}{E_{\text{potential,spring}}} = \frac{41.7\text{ ft-lbf}}{4.5\text{ ft-lbf}}$$
$$= 9.27$$

Rounding up to the nearest whole number, the minimum number of springs needed is 10.

The answer is (B).

Why Other Options Are Wrong

(A) This incorrect solution results from neglecting to square k and δ in the energy equations.

(C) This incorrect solution results from making the height 50 ft and the range 60 ft instead of vice versa.

(D) This solution results from neglecting to divide by g_c in the kinetic energy equation and neglecting to convert from inches to feet in the potential energy equation.

SOLUTION 38

The mean coil diameter is

$$d_{\text{coil}} = d_o - d_{\text{wire}} = 1.375\text{ in} - 0.125\text{ in}$$
$$= 1.25\text{ in}$$

The spring index is

$$C = \frac{d_{\text{coil}}}{d_{\text{wire}}} = \frac{1.25\text{ in}}{0.125\text{ in}}$$
$$= 10$$

The shear stress correction factor is

$$K_S = \frac{2C+1}{2C} = \frac{(2)(10)+1}{(2)(10)}$$
$$= 1.05$$

The Bergsträsser factor is

$$K_B = \frac{4C+2}{4C-3} = \frac{(4)(10)+2}{(4)(10)-3}$$
$$= 1.1351$$

The force amplitude is

$$F_a = \frac{F_{\max} - F_{\min}}{2} = \frac{5\text{ lbf} - 2\text{ lbf}}{2}$$
$$= 1.5\text{ lbf}$$

The mean force is

$$F_m = \frac{F_{\max} + F_{\min}}{2} = \frac{5\text{ lbf} + 2\text{ lbf}}{2}$$
$$= 3.5\text{ lbf}$$

The stress amplitude is

$$\tau_a = K_B\left(\frac{8F_a d_{\text{coil}}}{\pi d_{\text{wire}}^3}\right)$$
$$= (1.1351)\left(\frac{(8)(1.5\text{ lbf})(1.25\text{ in})}{\pi(0.125\text{ in})^3}\right)$$
$$= 2775\text{ lbf/in}^2$$

The mean stress is

$$\tau_m = K_S\left(\frac{8F_m d_{\text{coil}}}{\pi d_{\text{wire}}^3}\right)$$
$$= (1.05)\left(\frac{(8)(3.5\text{ lbf})(1.25\text{ in})}{\pi(0.125\text{ in})^3}\right)$$
$$= 5989\text{ lbf/in}^2$$

The maximum stress is

$$\tau_{\max} = \tau_a + \tau_m$$
$$= 2775\ \frac{\text{lbf}}{\text{in}^2} + 5989\ \frac{\text{lbf}}{\text{in}^2}$$
$$= 8764\text{ lbf/in}^2\quad(8800\text{ psi})$$

The answer is (D).

Why Other Options Are Wrong

(A) This incorrect solution results when the diameter of the wire is not cubed when calculating the stresses.

(B) This solution results when the maximum stress is incorrectly calculated as the average of the mean stress and stress amplitude instead of the sum.

(C) This solution results when the mean stress is mistaken for the maximum stress.

SOLUTION 39

The mean coil diameter of the spring is

$$d_{coil} = d_o - d_{wire} = 0.5 \text{ in} - 0.08 \text{ in}$$
$$= 0.42 \text{ in}$$

The shear modulus of stainless steel, taken from a table of material properties, is

$$G = 10 \times 10^6 \text{ psi}$$

The spring constant is

$$k = \frac{d_{wire}^4 G}{8 d_{coil}^3 N_{active}}$$
$$= \frac{(0.08 \text{ in})^4 \left(10 \times 10^6 \, \frac{\text{lbf}}{\text{in}^2}\right)}{(8)(0.42 \text{ in})^3 (30)}$$
$$= 23.0 \text{ lbf/in}$$

The density of stainless steel, also from a material property table, is

$$\rho = 0.280 \text{ lbm/in}^3$$

The mass of the active coils is

$$m = \frac{\pi^2 d_{wire}^2 d_{coil} N_{active} \rho}{4}$$
$$= \frac{\pi^2 (0.08 \text{ in})^2 (0.42 \text{ in})(30)\left(0.280 \, \frac{\text{lbm}}{\text{in}^3}\right)}{4}$$
$$= 0.055712 \text{ lbm}$$

The critical frequency is

$$f = \frac{1}{2}\sqrt{\frac{k g_c}{m}}$$
$$= \frac{1}{2}\sqrt{\frac{\left(23.0 \, \frac{\text{lbf}}{\text{in}}\right)\left(12 \, \frac{\text{lbf}}{\text{in}}\right)\left(32.2 \, \frac{\text{ft-lbm}}{\text{lbf-sec}^2}\right)}{0.055712 \text{ lbm}}}$$
$$= 199.7 \text{ Hz} \quad (200 \text{ Hz})$$

The answer is (C).

Why Other Options Are Wrong

(A) This incorrect solution includes π in the denominator of critical frequency calculation.

(B) This incorrect solution results when the outside diameter, instead of the mean coil diameter, is used.

(D) This incorrect solution results when the constant $1/2$ is left out of the frequency equation.

SOLUTION 40

The total shear stress on all the bolts is

$$\tau = \frac{F}{\left(\frac{\pi}{4}\right)d^2}$$
$$= \frac{25{,}000 \text{ lbf}}{\left(\frac{\pi}{4}\right)(0.5 \text{ in})^2}$$
$$= 127{,}300 \text{ lbf/in}^2$$

The number of bolts required to bear this stress is

$$N_{bolts} = \frac{\tau}{\tau_a} = \frac{127{,}300 \, \frac{\text{lbf}}{\text{in}^2}}{16{,}250 \, \frac{\text{lbf}}{\text{in}^2}}$$
$$= 7.83 \quad [\text{use } 8]$$

The cross-sectional area of the plates, at the plane running through the centers of the bolts, is

$$A = t(b - N_{bolts} d)$$
$$= (0.0625 \text{ in})(25 \text{ in} - (8)(0.5 \text{ in}))$$
$$= 1.313 \text{ in}^2$$

The tensile stress on the plate is

$$\sigma = \frac{F}{A}$$
$$= \frac{25{,}000 \text{ lbf}}{1.313 \text{ in}^2}$$
$$= 19{,}047 \text{ lbf/in}^2 \quad (19{,}000 \text{ psi})$$

The answer is (C).

Why Other Options Are Wrong

(A) This incorrect solution results when the tensile force of the bolts is computed as if for a pressure vessel.

(B) This incorrect solution results when the stress is computed simply as the force divided by the area of the plates.

(D) This incorrect solution results when the stress on the plates is assumed to be the force.

SOLUTION 41

Calculate the polar moment of inertia for each spindle.

$$J = \frac{\pi r^4}{2}$$

$$J_\text{I} = \frac{\pi \left(\frac{0.88 \text{ in}}{2}\right)^4}{2} = 0.0589 \text{ in}^4$$

$$J_\text{II} = \frac{\pi \left(\frac{1.25 \text{ in}}{2}\right)^4}{2} = 0.239 \text{ in}^4$$

$$J_\text{III} = \frac{\pi \left(\frac{1.00 \text{ in}}{2}\right)^4}{2} = 0.0982 \text{ in}^4$$

$$J_\text{IV} = \frac{\pi \left(\frac{1.12 \text{ in}}{2}\right)^4}{2} = 0.154 \text{ in}^4$$

Calculate the maximum torque for each spindle.

$$T = \frac{\tau J}{r}$$

$$T_\text{I} = \frac{\left(2500 \ \frac{\text{lbf}}{\text{in}^2}\right)(0.0589 \text{ in}^4)}{\frac{0.88 \text{ in}}{2}} = 335 \text{ in-lbf}$$

$$T_\text{II} = \frac{\left(2500 \ \frac{\text{lbf}}{\text{in}^2}\right)(0.239 \text{ in}^4)}{\frac{1.25 \text{ in}}{2}} = 956 \text{ in-lbf}$$

$$T_\text{III} = \frac{\left(2500 \ \frac{\text{lbf}}{\text{in}^2}\right)(0.0982 \text{ in}^4)}{\frac{1.00 \text{ in}}{2}} = 491 \text{ in-lbf}$$

$$T_\text{IV} = \frac{\left(2500 \ \frac{\text{lbf}}{\text{in}^2}\right)(0.154 \text{ in}^4)}{\frac{1.12 \text{ in}}{2}} = 688 \text{ in-lbf}$$

Calculate the maximum speed achieved by each spindle. The relationship between torque (in in-lbf) and power (in horsepower) is

$$T_\text{in-lbf} = \frac{63{,}025 P_\text{hp}}{n_\text{rpm}}$$

$$n_\text{rpm} = \frac{63{,}025 P_\text{hp}}{T_\text{in-lbf}}$$

$$n_\text{I,rpm} = \frac{(63{,}025)(19 \text{ hp})}{335 \text{ in-lbf}} = 3575 \text{ rpm}$$

$$n_\text{II,rpm} = \frac{(63{,}025)(22 \text{ hp})}{956 \text{ in-lbf}} = 1450 \text{ rpm}$$

$$n_\text{III,rpm} = \frac{(63{,}025)(28 \text{ hp})}{491 \text{ in-lbf}} = 3594 \text{ rpm}$$

$$n_\text{IV,rpm} = \frac{(63{,}025)(25 \text{ hp})}{688 \text{ in-lbf}} = 2290 \text{ rpm}$$

Model III has the fastest spindle speed, 3594 rpm.

The answer is (C).

Why Other Options Are Wrong

(A) This incorrect answer results when the model with the smallest diameter spindle is chosen. Although a smaller diameter leads to a greater speed, this is not quite enough to offset the model's low power.

(B) This incorrect answer results when the model with the highest torque is chosen. Increasing the torque leads to a slower spindle speed, not a greater one.

(D) This answer is incorrect because model IV's speed of 2290 rpm is lower than those of both model I and model III.

SOLUTION 42

The radius of the old shaft is

$$r = \frac{d_{\text{old}}}{2} = \frac{1.5 \text{ in}}{2} = 0.75 \text{ in}$$

The moment of inertia of the old shaft is

$$I = \frac{\pi r^4}{4} = \frac{\pi (0.75 \text{ in})^4}{4} = 0.249 \text{ in}^4$$

The modulus of elasticity for steel is 30×10^6 psi. The maximum deflection of the old shaft is

$$\delta = \frac{5wL^4}{384EI}$$

$$= \frac{(5)\left(0.9 \, \frac{\text{lbf}}{\text{in}}\right)(48 \text{ in})^4}{(384)\left(30 \times 10^6 \, \frac{\text{lbf}}{\text{in}^2}\right)(0.249 \text{ in}^4)}$$

$$= 0.00833 \text{ in}$$

The critical frequency of the old shaft is

$$f_{\text{old}} = \frac{1}{2\pi}\sqrt{\frac{g}{\delta}}$$

$$= \frac{1}{2\pi}\sqrt{\frac{32.2 \, \frac{\text{ft}}{\text{sec}^2}}{(0.00833 \text{ in})\left(\frac{1 \text{ ft}}{12 \text{ in}}\right)}}$$

$$= 34.278 \text{ Hz}$$

The new frequency required is

$$f_{\text{new}} = f_{\text{old}} + 10 \text{ Hz}$$
$$= 34.278 \text{ Hz} + 10 \text{ Hz}$$
$$= 44.278 \text{ Hz}$$

The new diameter required is

$$d_{\text{new}} = d_{\text{old}}\sqrt{\frac{f_{\text{new}}}{f_{\text{old}}}}$$

$$= 1.5 \text{ in}\sqrt{\frac{44.278 \text{ Hz}}{34.278 \text{ Hz}}}$$

$$= 1.705 \text{ in} \quad (1.7 \text{ in})$$

The answer is (B).

Why Other Options Are Wrong

(A) This incorrect solution results when diameter is used instead of radius to calculate the moment of inertia.

(C) This incorrect solution results when the square of the ratio of the frequencies is neglected for the calculation of the new shaft diameter.

(D) This incorrect solution results when the conversion from feet to inches is neglected during the calculation of the old frequency.

SOLUTION 43

The torque applied to the shaft by the motor is found by the formula

$$T_{\text{in-lbf}} = \frac{63{,}025 P_{\text{hp}}}{n_{\text{rpm}}}$$

$$= \frac{(63{,}025)(25 \text{ hp})}{5000 \text{ rpm}}$$

$$= 315.125 \text{ in-lbf}$$

The polar moment of inertia for the hollow shaft is

$$J = \frac{\pi}{2}(r_o^4 - r_i^4)$$

$$= \left(\frac{\pi}{2}\right)\left(\left(\frac{2.5 \text{ in}}{2}\right)^4 - \left(\frac{1.5 \text{ in}}{2}\right)^4\right)$$

$$= 3.34 \text{ in}^4$$

The shear stress at the outer surface of the shaft is

$$\tau = \frac{T r_o}{J}$$

$$= \frac{(315.125 \text{ in-lbf})(1.25 \text{ in})}{3.34 \text{ in}^4}$$

$$= 118 \text{ lbf/in}^2 \quad (120 \text{ psi})$$

The answer is (C).

Why Other Options Are Wrong

(A) This incorrect solution results when the radius is not quadrupled when calculating the polar moment of inertia.

(B) This incorrect solution results when diameter is used instead of radius.

(D) This incorrect solution results when the inside diameter is neglected.

SOLUTION 44

The torsional spring constant for each step of the shaft is found by the formula

$$k_r = \frac{\pi d^4 G}{32L}$$

$$k_{r,0.625} = \frac{\pi (0.625 \text{ in})^4 \left(11.5 \times 10^6 \, \frac{\text{lbf}}{\text{in}^2}\right)}{(32)(6 \text{ in})}$$
$$= 28{,}712 \text{ in-lbf}$$

$$k_{r,0.875} = \frac{\pi (0.875 \text{ in})^4 \left(11.5 \times 10^6 \, \frac{\text{lbf}}{\text{in}^2}\right)}{(32)(12 \text{ in})}$$
$$= 55150 \text{ in-lbf}$$

$$k_{r,0.500} = \frac{\pi (0.500 \text{ in})^4 \left(11.5 \times 10^6 \, \frac{\text{lbf}}{\text{in}^2}\right)}{(32)(8 \text{ in})}$$
$$= 8820 \text{ in-lbf}$$

$$k_{r,0.375} = \frac{\pi (0.375 \text{ in})^4 \left(11.5 \times 10^6 \, \frac{\text{lbf}}{\text{in}^2}\right)}{(32)(4 \text{ in})}$$
$$= 5582 \text{ in-lbf}$$

The reciprocal of the equivalent torsional spring constant for the entire shaft equals the sum of the reciprocals of the constants for its sections.

$$\frac{1}{k_{r,\text{eq}}} = \frac{1}{k_{r,0.625}} + \frac{1}{k_{r,0.875}} + \frac{1}{k_{r,0.500}} + \frac{1}{k_{r,0.375}}$$
$$= \frac{1}{28{,}712 \text{ in-lbf}} + \frac{1}{55{,}150 \text{ in-lbf}}$$
$$\quad + \frac{1}{8820 \text{ in-lbf}} + \frac{1}{5582 \text{ in-lbf}}$$
$$= \frac{1}{2894 \text{ in-lbf}}$$
$$k_{r,\text{eq}} = 2894 \text{ in-lbf} \quad (2900 \text{ in-lbf})$$

The answer is (C).

Why Other Options Are Wrong

(A) This incorrect solution results when the radius is used instead of the diameter.

(B) This incorrect solution results when the overall length is used to calculate each spring constant.

(D) This incorrect solution results when the equivalent spring constant is calculated in series instead of in parallel.

SOLUTION 45

Calculate head loss using the Darcy equation.

$$h_f = \frac{fL\text{v}^2}{2dg}$$

$$= \frac{(0.03)(32 \text{ ft})\left(20 \, \frac{\text{ft}}{\text{sec}}\right)^2}{(2)(3 \text{ in})\left(\frac{1 \text{ ft}}{12 \text{ in}}\right)\left(32.2 \, \frac{\text{ft}}{\text{sec}^2}\right)}$$
$$= 23.9 \text{ ft} \quad (24 \text{ ft})$$

The answer is (C).

Why Other Options Are Wrong

(A) This incorrect answer results from neglecting to square the velocity in the Darcy equation.

(B) This incorrect answer results from merely multiplying together the friction factor, length, and velocity.

(D) This answer results from remembering the Darcy equation incorrectly as $fL\text{v}/d$.

SOLUTION 46

The tensile stress area of a ¼-20 UNC threaded rod is found in thread tables to be

$$A = 0.0318 \text{ in}^2$$

The modulus of elasticity for steel is

$$E = 30 \times 10^6 \text{ psi}$$

The spring index of the bolt can be calculated from the formula

$$k = \frac{AE}{L}$$

$$= \frac{(0.0318 \text{ in}^2)\left(30 \times 10^6 \, \frac{\text{lbf}}{\text{in}^2}\right)}{12 \text{ in}}$$
$$= 79{,}500 \text{ lbf/in}$$

The change in length of the bolt from preloading can be calculated from the formula

$$k = \frac{F}{\Delta L}$$

$$\Delta L = \frac{F}{k}$$

$$= \frac{1200 \text{ lbf}}{79{,}500 \, \frac{\text{lbf}}{\text{in}}}$$

$$= 0.0151 \text{ in}$$

The degrees of turn required to preload the bolt is

$$\theta = (\text{turns per inch}) \Delta L$$

$$= \left(20 \, \frac{\text{rev}}{\text{in}}\right)\left(360 \, \frac{\text{deg}}{\text{rev}}\right)(0.0151 \text{ in})$$

$$= 108.72° \quad (100°)$$

The answer is (D).

Why Other Options Are Wrong

(A) This incorrect solution results when the threads per inch ratio is inverted when calculating the degrees of turn.

(B) This incorrect solution results when the threads per inch ratio is omitted when calculating the degrees of turn.

(C) This incorrect solution results when the tensile stress area is calculated using $1/4$ in as the diameter.

SOLUTION 47

The distance from the centroid of the bolt pattern to the most critical fastener is

$$r = \sqrt{x^2 + y^2} = \sqrt{\left(\frac{10 \text{ in}}{2}\right)^2 + \left(\frac{12 \text{ in}}{2}\right)^2}$$

$$= 7.810 \text{ in}$$

The area of each bolt is

$$A = \frac{\pi d^2}{4} = \left(\frac{\pi}{4}\right)(0.5 \text{ in})^2$$

$$= 0.1963 \text{ in}^2$$

The torsional resistance (or polar moment of inertia) of the fastener group is

$$J = nr^2 A = (4)(7.81 \text{ in})^2 (0.1963 \text{ in}^2)$$

$$= 47.89 \text{ in}^4$$

The eccentric distance for the bolt pattern is the distance from the centroid of the bolt pattern to the point load.

$$e = d + \frac{1}{2}x = 36 \text{ in} + \left(\frac{1}{2}\right)(10 \text{ in})$$

$$= 41 \text{ in}$$

The torsional shear stress of the fasteners is

$$\tau_t = \frac{Fer}{J} = \frac{(500 \text{ lbf})(41 \text{ in})(7.81 \text{ in})}{47.89 \text{ in}^4}$$

$$= 3343 \text{ lbf/in}^2$$

The angle is

$$\theta = \arctan \frac{x}{y} = \arctan \frac{10 \text{ in}}{12 \text{ in}}$$

$$= 39.81°$$

The horizontal shear stress is

$$\tau_x = (\cos \theta) \tau_t$$

$$= (\cos 39.81°)\left(3343 \, \frac{\text{lbf}}{\text{in}^2}\right)$$

$$= 2568 \text{ lbf/in}^2$$

The vertical shear stress is

$$\tau_y = (\sin \theta) \tau_t$$

$$= (\sin 39.81°)\left(3343 \, \frac{\text{lbf}}{\text{in}^2}\right)$$

$$= 2140 \text{ lbf/in}^2$$

The direct vertical shear is

$$\tau_v = \frac{F}{nA} = \frac{500 \text{ lbf}}{(4)(0.1963 \text{ in}^2)}$$

$$= 636.8 \text{ lbf/in}^2$$

The stress in the most critical fastener is

$$\tau = \sqrt{\tau_x^2 + (\tau_y + \tau_v)^2}$$

$$= \sqrt{\left(2568 \, \frac{\text{lbf}}{\text{in}^2}\right)^2 + \left(2140 \, \frac{\text{lbf}}{\text{in}^2} + 636.8 \, \frac{\text{lbf}}{\text{in}^2}\right)^2}$$

$$= 3782 \text{ lbf/in}^2 \quad (3800 \text{ psi})$$

The answer is (C).

Why Other Options Are Wrong

(A) This incorrect solution results when the stress in the most critical fastener is assumed to be the same as the direct vertical shear.

(B) This incorrect solution results when the stress in the most critical fastener is assumed to be the same as the torsional shear stress.

(D) This incorrect solution results when the horizontal and vertical stresses are reversed.

SOLUTION 48

The number of connectors, N, is 2. The bearing area is

$$A_b = tdN$$
$$= (0.025 \text{ in})(0.188 \text{ in})(2)$$
$$= 0.0094 \text{ in}^2$$

The sheet bearing failure load is

$$F_b = A_b S_c$$
$$= (0.0094 \text{ in}^2)\left(20{,}000 \, \frac{\text{lbf}}{\text{in}^2}\right)$$
$$= 188 \text{ lbf}$$

The rivet shear area is

$$A_{r,s} = \frac{\pi}{4} d^2 N$$
$$= \left(\frac{\pi}{4}\right)(0.188 \text{ in})^2 (2)$$
$$= 0.05552 \text{ in}^2$$

The rivet shear failure load is

$$F_{r,s} = A_{r,s} S_s$$
$$= (0.05552 \text{ in}^2)\left(8500 \, \frac{\text{lbf}}{\text{in}^2}\right)$$
$$= 472 \text{ lbf}$$

The sheet shear area is

$$A_{s,s} = N 2 t \left(e - \frac{d}{2}\right)$$
$$= (2)(2)(0.025 \text{ in})\left(0.3 \text{ in} - \frac{0.188 \text{ in}}{2}\right)$$
$$= 0.0206 \text{ in}^2$$

The sheet shear failure load is

$$F_{s,s} = A_{s,s} S_s$$
$$= (0.0206 \text{ in}^2)\left(8500 \, \frac{\text{lbf}}{\text{in}^2}\right)$$
$$= 175 \text{ lbf}$$

The width of the sheet is

$$b = 1.0 \text{ in} + (2)(0.3 \text{ in})$$
$$= 1.6 \text{ in}$$

The sheet tensile area is

$$A_t = (b - 2d)t$$
$$= (1.6 \text{ in} - (2)(0.188 \text{ in}))(0.025 \text{ in})$$
$$= 0.0306 \text{ in}^2$$

The sheet tension failure load is

$$F_t = A_t S_t$$
$$= (0.0306 \text{ in}^2)\left(10{,}000 \, \frac{\text{lbf}}{\text{in}^2}\right)$$
$$= 306 \text{ lbf}$$

The smallest failure load is 175 lbf (180 lbf) in sheet bearing failure.

The answer is (A).

Why Other Options Are Wrong

(B) This incorrect solution results when the sheet shearing failure load is calculated as the only failure mode.

(C) This incorrect solution results when the sheet tension failure load is calculated as the only failure mode.

(D) This incorrect solution results when the rivet shear failure load is calculated as the only failure mode.

SOLUTION 49

I. In a clamped connection, friction between the pieces is used to resist all of the movement of the pieces, and the connectors are never in a state of shear. (If the connection is loose, the connector will be placed into shear.)

II. Although the applied load is resisted by tension in the threaded member, friction between the threaded member is used to retain the nut.

III. Set screws depend on compression to develop the clamping force. Resistance to motion is the holding power due to the frictional resistance.

IV. In a press fit, some or all of the torsional resistance against rotation is provided by the friction between the shaft and hub. If a press fit is used, the pin often serves another purpose, such as ensuring alignment or indicating integrity.

Items II and III show connections that depend on friction for success. Items I and IV show connections that depend on position.

The answer is (D).

Why Other Options Are Wrong

(A) This answer is incorrect. Connection types III and IV are omitted.

(B) This answer is incorrect. Connection types II and IV are omitted.

(C) This answer is incorrect. Connection type IV is omitted.

SOLUTION 50

The symbol given in option A indicates a *gas tungsten arc* fillet weld with ½ in leg on both the arrow side and the back side, welded all around.

The symbol given in option B indicates a *gas metal arc* fillet weld with ½ in leg on the arrow side, welded all around.

The symbol given in option C indicates a *plasma arc* groove weld with a 75° groove angle on the arrow side and a 90° groove angle on the back side, welded at the indicated edge.

The symbol given in option D indicates an *electron beam* fillet spot weld 2 in long and welded on 7 in centers all around.

The four indicated processes are all suitable for joining aluminum. The weld is to be on the arrow side only, because it is not possible to weld the back side of the box beam joint as described, so symbol A is wrong. A fillet weld, not a groove weld, is indicated, eliminating symbol C. All four edges are to be completely welded, eliminating symbol D, which shows an intermittent weld.

The answer is (B).

Why Other Options Are Wrong

(A) This symbol is incorrect because the weld is to be on the arrow side only.

(C) This symbol is incorrect because it shows a groove weld when a fillet weld is called for.

(D) This symbol is incorrect because it shows an intermittent weld when a continuous weld is required.

SOLUTION 51

The moment of inertia is

$$I = \frac{\pi r^4}{4} = \frac{\pi \left(\dfrac{d}{2}\right)^4}{4}$$

$$= \frac{\pi \left(\dfrac{1.375 \text{ in}}{2}\right)^4}{4}$$

$$= 0.1755 \text{ in}^4$$

The modulus of elasticity of steel is

$$E = 30 \times 10^6 \text{ psi}$$

The static deflection at the mass is

$$\delta_{\text{st}} = \frac{Fa^2b^2}{3EIL}$$

$$= \frac{(250 \text{ lbf})(9 \text{ in})^2(30 \text{ in} - 9 \text{ in})^2}{(3)\left(30 \times 10^6 \dfrac{\text{lbf}}{\text{in}^2}\right)(0.1755 \text{ in}^4)(30 \text{ in})}$$

$$= 0.01885 \text{ in}$$

The critical speed is

$$f = \frac{1}{2\pi}\sqrt{\frac{g}{\delta_{\text{st}}}}$$

$$= \frac{1}{2\pi}\sqrt{\frac{\left(32.2 \dfrac{\text{ft}}{\text{sec}^2}\right)\left(12 \dfrac{\text{in}}{\text{ft}}\right)}{0.01885 \text{ in}}}$$

$$= 22.8 \text{ sec}^{-1} \quad (22.8 \text{ Hz})$$

The operating speed is half the critical speed.

$$f_{\text{op}} = \frac{22.8 \text{ Hz}}{2}$$

$$= 11.4 \text{ Hz}$$

$$\omega = 2\pi f_{\text{op}}$$

$$= 2\pi(11.4 \text{ Hz})\left(\frac{1 \text{ sec}^{-1}}{1 \text{ Hz}}\right)$$

$$= 71.6 \text{ rad/sec} \quad (72 \text{ rad/sec})$$

The answer is (B).

Why Other Options Are Wrong

(A) This incorrect solution results when the deflection of the shaft is calculated with a center load deflection formula.

(C) This incorrect solution results when the diameter, instead of the radius, is used for calculations.

(D) This incorrect solution results when the maximum deflection of the shaft, instead of the deflection of the shaft at the mass, is used.

SOLUTION 52

The forcing frequency is

$$f_f = \left(100 \ \frac{\text{rev}}{\text{min}}\right)\left(\frac{1 \ \text{min}}{60 \ \text{sec}}\right)$$
$$= 1.667 \ \text{Hz}$$

The static deflection can be found from the equation

$$f_f = f_n = \frac{1}{2\pi}\sqrt{\frac{g}{\delta_{\text{st}}}}$$

$$\delta_{\text{st}} = \frac{g}{(2\pi f_f)^2}$$

$$= \frac{32.2 \ \frac{\text{ft}}{\text{sec}^2}}{(2\pi(1.667 \ \text{Hz}))^2}$$

$$= 0.2935 \ \text{ft}$$

The spring stiffness is

$$k = \frac{mg}{\delta_{\text{st}} g_c}$$

$$= \frac{(5 \ \text{lbm})\left(32.2 \ \frac{\text{ft}}{\text{sec}^2}\right)}{(0.2935 \ \text{ft})\left(32.2 \ \frac{\text{ft-lbm}}{\text{lbf-sec}^2}\right)}$$

$$= 17.04 \ \text{lbf/ft} \quad (17 \ \text{lbf/ft})$$

The answer is (C).

Why Other Options Are Wrong

(A) This incorrect answer results if the $1/2\pi$ term is omitted in calculating the natural frequency.

(B) This incorrect answer results if the mass (in lbm) is divided only by g_c without also multiplying by g.

(D) This incorrect answer results if the mass is multiplied by g without dividing by g_c.

SOLUTION 53

The static deflection is

$$\delta_{\text{st}} = \frac{mg}{k g_c}$$

$$= \frac{(7520 \ \text{lbm})\left(32.2 \ \frac{\text{ft}}{\text{sec}^2}\right)}{\left(30{,}000 \ \frac{\text{lbf}}{\text{in}}\right)\left(32.2 \ \frac{\text{ft-lbm}}{\text{lbf-sec}^2}\right)}$$

$$= 0.25 \ \text{in}$$

The undamped natural frequency is

$$f_n = \left(\frac{1}{2\pi}\right)\sqrt{\frac{g}{\delta_{\text{st}}}}$$

$$= \frac{1}{2\pi}\sqrt{\frac{32.2 \ \frac{\text{ft}}{\text{sec}^2}}{(0.25 \ \text{in})\left(\frac{1 \ \text{ft}}{12 \ \text{in}}\right)}}$$

$$= 6.26 \ \text{Hz}$$

The forcing frequency is

$$f_f = \left(70 \ \frac{\text{rev}}{\text{min}}\right)\left(\frac{1 \ \text{min}}{60 \ \text{sec}}\right)$$
$$= 1.1667 \ \text{rev/sec} \quad (1.1667 \ \text{Hz})$$

The angular forcing frequency is

$$\omega_f = 2\pi f_f$$
$$= 2\pi(1.667 \ \text{Hz})$$
$$= 7.3304 \ \text{rad/sec}$$

The force caused by the rotating eccentric mass is

$$F_f = \frac{m\omega_f^2 r}{g_c}$$

$$= \frac{(20 \ \text{lbm})(7.3304 \ \text{Hz})^2 \left(\frac{90 \ \text{in}}{2}\right)\left(\frac{1 \ \text{ft}}{12 \ \text{in}}\right)}{32.2 \ \frac{\text{ft-lbm}}{\text{lbf-sec}^2}}$$

$$= 125.16 \ \text{lbf}$$

The ratio of frequencies is

$$r = \frac{f_f}{f_n}$$
$$= \frac{1.1667 \text{ Hz}}{6.26 \text{ Hz}}$$
$$= 0.186$$

The magnification factor is

$$\text{MF} = \left| \frac{1}{\sqrt{(1-r^2)^2 + (2\zeta r)^2}} \right|$$
$$= \left| \frac{1}{\sqrt{\left(1-(0.186)^2\right)^2 + \left((2)(0.05)(0.186)\right)^2}} \right|$$
$$= 1.0356$$

The transmissibility is

$$\text{TR} = (\text{MF})\sqrt{1 + (2r\zeta)^2}$$
$$= (1.0356)\sqrt{1 + ((2)(0.186)(0.05))^2}$$
$$= 1.0358$$

The transmitted force is

$$F_t = (\text{TR})F_f$$
$$= (1.0358)(125.16 \text{ lbf})$$
$$= 129.6 \text{ lbf} \quad (130 \text{ lbf})$$

The answer is (B).

Why Other Options Are Wrong

(A) This incorrect solution results when the transmissibility is assumed to be the remaining portion of the damping factor.

(C) This incorrect solution results when the first frequency ratio in the denominator of the magnification factor is not squared.

(D) This incorrect solution results when the mass of the reel is used to calculate the out-of-balance force as the transmitted force, and the decimal point is adjusted to bring the answer into a reasonable range.

SOLUTION 54

The mass impact velocity is

$$v = \sqrt{2gh}$$
$$= \sqrt{(2)\left(32.2 \frac{\text{ft}}{\text{sec}^2}\right)(24 \text{ in})\left(\frac{1 \text{ ft}}{12 \text{ in}}\right)}$$
$$= 11.349 \text{ ft/sec}$$

The angular frequency is

$$\omega = \frac{v}{\delta} = \frac{11.349 \frac{\text{ft}}{\text{sec}}}{(6 \text{ in})\left(\frac{1 \text{ ft}}{12 \text{ in}}\right)}$$
$$= 22.698 \text{ sec}^{-1}$$

The transmitted shock (acceleration) is

$$a = \omega v$$
$$= (22.698 \text{ sec}^{-1})\left(11.349 \frac{\text{ft}}{\text{sec}}\right)$$
$$= 257.6 \text{ ft/sec}^2$$

The required mass is

$$m = \frac{Fg_c}{a}$$
$$= \frac{(5000 \text{ lbf})\left(32.2 \frac{\text{ft-lbm}}{\text{lbf-sec}^2}\right)}{257.6 \frac{\text{ft}}{\text{sec}^2}}$$
$$= 625 \text{ lbm} \quad (630 \text{ lbm})$$

The answer is (C).

Why Other Options Are Wrong

(A) This incorrect solution results when the gravitational constant, g_c, is neglected.

(B) This incorrect solution results when a straight calculation is made using the force and the acceleration of gravity to calculate the mass, neglecting the height.

(D) This incorrect solution results when the natural frequency, instead of the angular frequency, is used and the 2π conversion factor is neglected.

SOLUTION 55

From the static balance,

$$\sum m_i r_i = 0 \text{ lbm-in}$$

$$m_{\text{binder}} r_{\text{binder}} + m_{\text{die}} r_{\text{die}}$$
$$- m_{\text{counterweight}} r_{\text{counterweight}}$$
$$= 0 \text{ lbm-in}$$

$$(15 \text{ lbm})(4 \text{ in}) - (10 \text{ lbm})(3 \text{ in})$$
$$- (m_{\text{counterweight}})(3 \text{ in})$$
$$= 0 \text{ lbm-in}$$
$$m_{\text{counterweight}} = 10 \text{ lbm}$$

The sum of the moments around the die must be zero, so

$$m_{\text{binder}} r_{\text{binder}} x_{\text{binder}}$$
$$- m_{\text{counterweight}} r_{\text{counterweight}} x_{\text{counterweight}}$$
$$= 0 \text{ lbm-in}^2$$

$$(15 \text{ lbm})(4 \text{ in})(8 \text{ in})$$
$$- (10 \text{ lbm})(3 \text{ in})(d + 8 \text{ in})$$
$$= 0 \text{ lbm-in}^2$$
$$d = 8.0 \text{ in}$$

The answer is (C).

Why Other Options Are Wrong

(A) This incorrect solution results when the sign convention is ignored in calculating the mass of the counterweight.

(B) This incorrect solution results when the eccentricity of each component is neglected while calculating the counterweight position.

(D) This incorrect solution results when the distance between the binder and the die is neglected while calculating the counterweight position.

SOLUTION 56

The mean stress is

$$\sigma_m = \sqrt{\sigma_{m1}^2 + \sigma_{m2}^2 - \sigma_{m1}\sigma_{m2}}$$

$$= \sqrt{\left(2500 \frac{\text{lbf}}{\text{in}^2}\right)^2 + \left(750 \frac{\text{lbf}}{\text{in}^2}\right)^2 - \left(2500 \frac{\text{lbf}}{\text{in}^2}\right)\left(750 \frac{\text{lbf}}{\text{in}^2}\right)}$$

$$= 2222 \text{ lbf/in}^2 \quad (2200 \text{ psi})$$

The alternating von Mises stress is

$$\sigma_{\text{alt}} = \sqrt{\sigma_{\text{alt1}}^2 + \sigma_{\text{alt2}}^2 - \sigma_{\text{alt1}}\sigma_{\text{alt2}}}$$

$$= \sqrt{\left(500 \frac{\text{lbf}}{\text{in}^2}\right)^2 + \left(1500 \frac{\text{lbf}}{\text{in}^2}\right)^2 - \left(500 \frac{\text{lbf}}{\text{in}^2}\right)\left(1500 \frac{\text{lbf}}{\text{in}^2}\right)}$$

$$= 1322 \text{ lbf/in}^2 \quad (1300 \text{ psi})$$

The answer is (C).

Why Other Options Are Wrong

(A) This incorrect solution results when the stresses are used as maximum and minimum stresses to find the alternating stress and averaged to find the mean stress.

(B) This incorrect solution results when the mean and alternating stresses are both found by averaging.

(D) This incorrect solution results when the mean and alternating stresses are both found by adding.

SOLUTION 57

Heat treatment is the process of changing the microstructure of the steel using temperature and time. Deterioration of structure and losses of hardness and magnetism are effects resulting from prolonging the austenitizing temperature soak. Options I, II, and IV are correct.

Carburization is the process of adding carbon to the outer surface of the steel during heat treatment to increase the case hardness. Option III is incorrect.

The answer is (C).

Why Other Options Are Wrong

(A) This answer is wrong. Carburization is not a result of exended exposure.

(B) This answer is wrong. Carburization is not a result of exended exposure.

(D) This answer is wrong. Loss of hardness is a result of extended exposure, and carburization is not.

SOLUTION 58

Given x repetitions of the pattern, the numbers of cycles at the first and second stress levels are

$$n_1 = 5x \text{ cycles/repetition}$$
$$n_2 = 10x \text{ cycles/repetition}$$

The numbers of cycles in the fatigue life at the first and second stress levels are

$$N_1 = 10^7 \text{ cycles}$$
$$N_2 = 10^6 \text{ cycles}$$

The number of repetitions of the pattern that will fulfill the failure criterion is found by the formula

$$\frac{n_1}{N_1} + \frac{n_2}{N_2} = C$$

$$\frac{5x \frac{\text{cycles}}{\text{repetition}}}{10^7 \text{ cycles}} + \frac{10x \frac{\text{cycles}}{\text{repetition}}}{10^6 \text{ cycles}} = 0.9$$

$$x = \frac{0.9}{\frac{5 \frac{\text{cycles}}{\text{repetition}}}{10^7 \text{ cycles}} + \frac{10 \frac{\text{cycles}}{\text{repetition}}}{10^6 \text{ cycles}}}$$

$$= 85{,}714 \text{ repetitions} \quad (0.86 \times 10^5 \text{ repetitions})$$

The answer is (A).

Why Other Options Are Wrong

(B) This incorrect solution results when x is given the value of 0.9, the formula is solved for C, and the solution is inverted to achieve a reasonable value.

(C) This solution results when the number of cycles at the second stress level is incorrectly thought to be five.

(D) This solution results when the formula is incorrectly thought to be

$$\frac{n_1 + n_2}{N_1 + N_2} = C$$

SOLUTION 59

Use the NTU method. First, find the thermal capacity rate for the cold fluid.

$$\eta_{\text{exchanger}} = \frac{C_{\text{hot}}(T_{\text{hot,in}} - T_{\text{hot,out}})}{C_{\text{cold}}(T_{\text{hot,in}} - T_{\text{cold,in}})}$$

$$C_{\text{cold}} = \frac{C_{\text{hot}}(T_{\text{hot,in}} - T_{\text{hot,out}})}{\eta_{\text{exchanger}}(T_{\text{hot,in}} - T_{\text{cold,in}})}$$

$$= \frac{\left(2.0 \frac{\text{Btu}}{\text{sec-°F}}\right)(50°\text{F})}{(0.90)(100°\text{F})}$$

$$= 1.11 \text{ Btu/sec-°F}$$

Use the thermal capacity rate and the specific heat of the cold fluid to calculate the mass flow rate.

$$\dot{m} = \frac{C_{\text{cold}}}{c_p}$$

$$= \frac{1.11 \frac{\text{Btu}}{\text{sec-°F}}}{1.0 \frac{\text{Btu}}{\text{lbm-°F}}}$$

$$= 1.11 \frac{\text{lbm}}{\text{sec}}$$

Calculate the hydraulic horsepower to pump the cold fluid.

$$\text{WHP} = \left(\frac{h_A \dot{m}}{550}\right)\left(\frac{g}{g_c}\right)$$

$$= \left(\frac{(5000 \text{ ft})\left(1.11 \frac{\text{lbm}}{\text{sec}}\right)}{550 \frac{\text{ft-lbf}}{\text{hp-sec}}}\right)\left(\frac{32.2 \frac{\text{ft}}{\text{sec}^2}}{32.2 \frac{\text{ft-lbm}}{\text{lbf-sec}^2}}\right)$$

$$= 10.09 \text{ hp}$$

Calculate the pump motor horsepower.

$$\text{BHP} = \frac{\text{WHP}}{\eta_{\text{pump}}}$$

$$= \frac{10.09 \text{ hp}}{0.95}$$

$$= 10.62 \text{ hp} \quad (11 \text{ hp})$$

The answer is (C).

Why Other Options Are Wrong

(A) This incorrect answer results when the efficiency of the heat exchanger is neglected or thought to be 100% in calculating the cold water mass flow rate.

(B) This incorrect answer results when the efficiency of the pump is neglected or thought to be 100%.

(D) This incorrect answer results when the efficiency of the heat exchanger is thought to be simply the hot water thermal capacity divided by the cold water thermal capacity.

SOLUTION 60

Calculate the heat generated by the steam plant.

$$H = \eta_{\text{boiler}} \dot{m}(\text{HHV})$$

$$= (0.90)\left(5 \ \frac{\text{ton}}{\text{day}}\right)\left(11{,}000 \ \frac{\text{Btu}}{\text{lbm}}\right)\left(2000 \ \frac{\text{lbf}}{\text{ton}}\right)$$

$$\times \left(\frac{32.2 \ \frac{\text{ft-lbm}}{\text{lbf-sec}^2}}{32.2 \ \frac{\text{ft}}{\text{sec}^2}}\right)\left(\frac{1 \ \text{day}}{24 \ \text{hr}}\right)$$

$$= 4{,}125{,}000 \ \text{Btu/hr}$$

Water exiting the condenser has the lowest temperature within the steam plant, and steam exiting the boiler has the highest temperature. Convert these temperatures to the Rankine scale, and calculate the Carnot efficiency.

$$\eta_{\text{Carnot}} = 1 - \frac{T_L}{T_H}$$

$$= 1 - \frac{100° \text{F} + 460}{1000° \text{F} + 460}$$

$$= 0.616 \quad (0.62)$$

Calculate the maximum theoretical power output.

$$Q_{\max} = \eta_{\text{Carnot}} H - Q_{\text{pump}}$$

$$= \eta_{\text{Carnot}} H - (0.05) Q_{\max}$$

$$= \frac{\eta_{\text{Carnot}} H}{1.05}$$

$$= \frac{(0.62)\left(4{,}125{,}000 \ \dfrac{\text{Btu}}{\text{hr}}\right)}{1.05}$$

$$= 2{,}435{,}714 \ \text{Btu/hr} \quad (2.4 \times 10^6 \ \text{Btu/hr})$$

The answer is (A).

Why Other Options Are Wrong

(B) This incorrect answer results from neglecting the boiler efficiency.

(C) This incorrect answer results from switching the high and low temperatures when calculating the Carnot efficiency.

(D) This incorrect answer results from neglecting the Carnot efficiency.

SOLUTION 61

A statistical process control chart can be prepared for the average of some process control variable. A process is in "control" as long as the plotted data remain within the control limits. The upper and lower control limits are $\bar{x} \pm 3\sigma$, or three times the standard deviation of all data points plotted on each side of the average.

11 of the last 12 data points are on one side of the average even though the points are within the control limits. This zone test indicates that a suspicious event has occurred. In this case the process average has shifted. Other zone tests include 12 of 14, 14 of 17, and 16 of 20 data points on one side of the average. Statements I and II are both correct.

A trend occurs when the data are plotted in a defined direction with a suspicious event. The data for this chart are steadily decreasing before reaching the shift in the average. Statement III is correct.

Bimodal data are indicated when a mixture pattern occurs. All data are plotted very closely to either control limit, but the data do not distribute well throughout the graph or near the average. Different machines, operators, or raw material lots can make this extreme greater than the variation within the process. Likewise, stratification occurs when almost all data points are near the average. This indicates either an error in control limit calculations or questionable sample selection techniques. Statement IV is incorrect.

The answer is (C).

Why Other Options Are Wrong

(A) This answer is incorrect. Statements II and III are both true, while statement IV is false.

(B) This answer lists the false statement, not the true ones.

(D) This answer is incorrect because it excludes statement I, which is true.

SOLUTION 62

The molecular weight of CO_2 can be found in reference tables, and is equal to 44 lbm/lbmol. The specific gas constant, R, is

$$R = \frac{R*}{\text{MW}}$$

$$= \frac{1545.33 \ \frac{\text{ft-lbf}}{\text{lbmol-°R}}}{44 \ \frac{\text{lbm}}{\text{lbmol}}}$$

$$= 35.12 \ \text{ft-lbf/lbm-°R}$$

The temperature is

$$T = 70° \text{F} + 460°$$

$$= 530°\text{R}$$

The volume of the cylinder is

$$V = \pi \left(\frac{d}{2}\right)^2 L$$

$$= \pi \left(\left(\frac{4 \text{ in}}{2}\right)\left(\frac{1 \text{ ft}}{12 \text{ in}}\right)\right)^2 (10 \text{ ft})$$

$$= 0.873 \text{ ft}^3$$

The pressure of carbon dioxide, treated as an ideal gas, within the cylinder is

$$p = \frac{mRT}{V}$$

$$= \frac{(10 \text{ lbm})\left(35.12 \ \frac{\text{ft-lbf}}{\text{lbm-}°\text{R}}\right)(530°\text{R})}{0.873 \text{ ft}^3} \left(\frac{1 \text{ ft}}{12 \text{ in}}\right)^2$$

$$= 1481 \text{ lbf/in}^2$$

Poisson's ratio for steel is

$$\nu = 0.3$$

The cylinder is made of SAE 1020 steel, which is a ductile material. Ductile material includes low-carbon steel (less than 0.30% carbon content), brass, bronze, and so on. Clavarino's equation applies to a thick-walled cylinder of ductile material that is closed at both ends.

$$t = \left(\frac{d}{2}\right)\left(\sqrt{\frac{S + (1 - 2\nu)p}{S - (1 + \nu)p}} - 1\right)$$

$$= \left(\frac{4 \text{ in}}{2}\right)$$

$$\times \left(\sqrt{\frac{5000 \ \frac{\text{lbf}}{\text{in}^2} + (1 - (2)(0.3))\left(1481 \ \frac{\text{lbf}}{\text{in}^2}\right)}{5000 \ \frac{\text{lbf}}{\text{in}^2} - (1 + 0.3)\left(1481 \ \frac{\text{lbf}}{\text{in}^2}\right)}} - 1\right)$$

$$= 0.697 \text{ in} \quad (0.70 \text{ in})$$

The answer is (C).

Why Other Options Are Wrong

(A) This incorrect solution results from not converting the temperature to Rankine.

(B) This incorrect solution results from incorrectly converting the temperature to Rankine by adding 273 instead of 460.

(D) This solution results from incorrectly assuming a MW of 28 for carbon dioxide. (The MW of carbon monoxide is 28.)

SOLUTION 63

Let x be the number of hours of operation. The equivalent uniform annual cost (EUAC) for plan A is

$$\text{EUAC}_\text{A} = (\$0.40)x$$

The mean time between failures (MTBF) is 3002 hr, and the mean time to failure (MTTF) is 3000 hr. Therefore, the mean time to repair (MTTR) is

$$\text{MTTR} = \text{MTBF} - \text{MTTF}$$
$$= 3002 \text{ hr} - 3000 \text{ hr}$$
$$= 2 \text{ hr}$$

The system will require 2 hr of repair time for every 3000 hr of operation. The maintenance cost per hour is

$$\text{MC} = \$0.25$$

The mean repair cost per hour of operation must be

$$\text{MRC} = \left(\frac{\text{hr of repair}}{\text{hr of operation}}\right)(\text{repair cost per hour})$$

$$= \left(\frac{2 \text{ hr}}{3000 \text{ hr}}\right)(\$50)$$

$$= \$0.33$$

The EUAC for plan B is

$$\text{EUAC}_\text{B} = \text{annual value of initial cost}$$
$$- \text{annual value of salvage}$$
$$+ \text{preventative maintenance}$$
$$+ \text{corrective maintenance}$$
$$= (\$10{,}000)(A/P,7\%,10)$$
$$- (\$1000)(A/F,7\%,10)$$
$$+ (\text{MC})x + (\text{MRC})x$$

The annual values are found from a factor table for 7% interest.

$$\text{EUAC}_\text{B} = (\$10{,}000)(0.1424) - (\$1000)(0.0724)$$
$$+ (\$0.25)x + (\$0.033)x$$
$$= \$1351.6 + (\$0.283)x$$

To determine the break-even point, find the value of x that makes EUAC_A and EUAC_B equal.

$$\text{EUAC}_\text{A} = \text{EUAC}_\text{B}$$
$$(\$0.40)x = \$1351.6 + (\$0.283)x$$
$$x = \frac{\$1351.6}{\$0.40 - \$0.283}$$
$$= 11{,}552 \text{ hr}$$

Because the system will be operating 3000 hr/yr, the break-even point can be estimated as

$$\text{break-even point} = \frac{\text{total hours}}{\text{hours per year}}$$
$$= \frac{11{,}552 \text{ hr}}{3000 \ \frac{\text{hr}}{\text{yr}}}$$
$$= 3.8506 \text{ yr} \quad (3.9 \text{ yr})$$

The answer is (B).

Why Other Options Are Wrong

(A) This incorrect answer results from neglecting the effect of the 7% interest rate on the annual values, and simply dividing the initial cost and salvage value by 10.

(C) This incorrect answer results from adding the salvage value to the costs instead of subtracting it.

(D) This answer results from incorrectly calculating the mean repair cost per hour of operation as $0.05.

SOLUTION 64

The present value of the initial costs is

$$P = \text{acquisition cost} + \text{reconfiguration cost}$$
$$= \$3000 + \$1000$$
$$= \$4000$$

The future value of the plant is

$$F = \text{salvage value} + \text{reconfiguration value}$$
$$= \$1000 + \$1000$$
$$= \$2000$$

The annual cost to maintain the plant is

$$A = \$400$$

The discount factor for the present value, P, is

$$(A/P,5\%,5) = 0.2310$$

The discount factor for the salvage value is

$$(A/F,5\%,5) = 0.1810$$

Therefore, the EUAC is

$$\text{EUAC} = A + P(A/P,5\%,5) + F(A/F,5\%,5)$$
$$= \$400 + (\$4000)(0.2310) - (\$2000)(\$0.1810)$$
$$= \$962 \quad (\$960)$$

The answer is (C).

Why Other Options Are Wrong

(A) This incorrect answer results from omitting the annual maintenance cost of $400.

(B) This incorrect answer results from using $(A/F, 5\%, 5)$ as the discount factor for the present value.

(D) This incorrect answer results from using $(P/A, 5\%, 5)$ as the discount factor for the present value.

SOLUTION 65

Equate the spring force and the force developed by the column of liquid.

$$F_{\text{spring}} = F_{\text{column}}$$
$$k\delta = \gamma h A$$

Solve for the height of the liquid column.

$$h = \frac{k\delta}{\gamma A}$$
$$= \frac{\left(1000 \ \frac{\text{lbf}}{\text{in}}\right)(3 \text{ in})}{\left(62.4 \ \frac{\text{lbf}}{\text{ft}^3}\right)(1 \text{ ft}^2)}$$
$$= 48.1 \text{ ft} \quad (48 \text{ ft})$$

The answer is (D).

Why Other Options Are Wrong

(A) This answer results when units are canceled incorrectly and the final result is thought to be in inches instead of feet.

(B) This incorrect answer results when the spring force is calculated by dividing the spring constant by the deflection instead of multiplying the spring constant and deflection.

(C) This incorrect answer results when the spring deflection is omitted from the calculation.

SOLUTION 66

Start by creating a table that itemizes the parameters.

	option			
	A	B	C	D
process	old	new	old	old
material	old	old	new	old
price	old	old	—	new
process cost, PC ($/hr)	120	210	120	120
material cost, MC ($/lbm)	0.65	0.65	0.53	0.55
qualification cost ($)	0	2 million	2 million	0
contract fee ($)	0	0	0	100,000
material consumption rate, MCR (lbm/hr)	2500	2250	2500	2500

The material consumption rate for the modified process is

$$\text{MCR}_{\text{new}} = (\text{MCR}_{\text{old}})(90\%) = \left(2500 \ \frac{\text{lbm}}{\text{hr}}\right)(0.90)$$
$$= 2250 \ \text{lbm/hr}$$

The total process cost is

$$\text{PC}_{5 \ \text{yr}} = \eta t (\text{PC}_{\text{hr}})$$

For the old process, this is

$$\text{PC}_{5 \ \text{yr,old}} = (0.75)(5 \ \text{yr})\left(24 \ \frac{\text{hr}}{\text{day}}\right)\left(365 \ \frac{\text{day}}{\text{yr}}\right)$$
$$\times \left(\frac{\$120}{\text{hr}}\right)$$
$$= \$3{,}942{,}000$$

For the modified process, this is

$$\text{PC}_{5 \ \text{yr,new}} = (0.75)(5 \ \text{yr})\left(24 \ \frac{\text{hr}}{\text{day}}\right)\left(365 \ \frac{\text{day}}{\text{yr}}\right)$$
$$\times \left(\frac{\$210}{\text{hr}}\right)$$
$$= \$6{,}898{,}500$$

The total material cost is

$$\text{MC}_{5 \ \text{yr}} = \eta t (\text{MCR})(\text{MC}_{\text{lbm}})$$

For option A, this is

$$\text{MC}_{5 \ \text{yr,A}} = (0.75)(5 \ \text{yr})\left(24 \ \frac{\text{hr}}{\text{day}}\right)\left(365 \ \frac{\text{day}}{\text{yr}}\right)$$
$$\times \left(2500 \ \frac{\text{lbm}}{\text{hr}}\right)\left(\frac{\$0.65}{\text{lbm}}\right)$$
$$= \$53{,}381{,}250$$

For option B, this is

$$\text{MC}_{5 \ \text{yr,B}} = (0.75)(5 \ \text{yr})\left(24 \ \frac{\text{hr}}{\text{day}}\right)\left(365 \ \frac{\text{day}}{\text{yr}}\right)$$
$$\times \left(2250 \ \frac{\text{lbm}}{\text{hr}}\right)\left(\frac{\$0.65}{\text{lbm}}\right)$$
$$= \$48{,}043{,}125$$

For option C, this is

$$\text{MC}_{5 \ \text{yr,C}} = (0.75)(5 \ \text{yr})\left(24 \ \frac{\text{hr}}{\text{day}}\right)\left(365 \ \frac{\text{day}}{\text{yr}}\right)$$
$$\times \left(2500 \ \frac{\text{lbm}}{\text{hr}}\right)\left(\frac{\$0.53}{\text{lbm}}\right)$$
$$= \$43{,}526{,}250$$

For option D, this is

$$\text{MC}_{5 \ \text{yr,D}} = (0.75)(5 \ \text{yr})\left(24 \ \frac{\text{hr}}{\text{day}}\right)\left(365 \ \frac{\text{day}}{\text{yr}}\right)$$
$$\times \left(2500 \ \frac{\text{lbm}}{\text{hr}}\right)\left(\frac{\$0.55}{\text{lbm}}\right)$$
$$= \$45{,}168{,}750$$

Create a table of options for summing and comparison.

	option			
	A	B	C	D
total process cost ($)	3,942,000	6,898,500	3,942,000	3,942,000
total material cost ($)	53,381,250	48,043,125	43,526,250	45,168,750
qualification cost ($)	0	2,000,000	2,000,000	0
contract fee ($)	0	0	0	100,000
total costs ($)	57,323,250	56,941,625	49,468,250	49,210,750

The answer is (D).

Why Other Options Are Wrong

(A) This incorrect solution results when only the process cost and the qualification and contract costs are considered.

(B) This incorrect solution results when only the material consumption rate is considered.

(C) This incorrect solution results when qualification and contract costs are neglected.